An Important Message to Our Readers

Gas-Core Nuclear Rocket Design

A Master's Thesis

by

David Hitchcock

First Edition

Front Cover Art   NASA

Back Cover Art   NASA

Publisher's Cataloging-in-Publication data

Hitchcock, David, 1956-
   Gas-core nuclear rocket design : a Master's thesis / by David Hitchcock.
   p. cm.
   1st edition.
   ISBN 978-1-4357-5032-6
   Includes bibliographical references.
1. Space vehicles --Nuclear power plants. 2. Space vehicles--Propulsion systems. 3. Nuclear rockets. 4. Space flight. I. Title.

TL1102.N8 H58 2008
629.47/443 19--dc 22   2008907136

Dedication

This book is dedicated to all the rocket propulsion engineers and scientists of the world, who toil endlessly to make space exploration a reality.

This book is also dedicated with love and appreciation to my wife, Sylvia and my daughter, Jennifer.

About the Author

David Hitchcock has degrees in physics and engineering, and has worked as a computer consultant on such diverse projects as the MX missile, the Milstar satellite program, advanced capability torpedoes and jet engine computer simulations for NASA.

Mr. Hitchcock is also the author of *Patent Searching Made Easy*, currently in its fourth edition.

CONTENTS

TABLE OF FIGURES

iv

LIST OF TABLES

1.0   INTRODUCTION

The Gas Core Nuclear Rocket has the potential for significant improvement in propulsion capability over Solid Core Nuclear and Chemical Rockets. In order to realize this potential, however, it is necessary to overcome certain design problems inherent with the Gas-Core. Among the most important of these are: Nuclear Fuel Containment and Nuclear Criticality. Although there are several Gas-Core design concepts, only three of them can be considered mature (in the sense of analytical and experimental probing short mainly of fission power tests); Colloid Core, Nuclear Light-Bulb, and Coaxial Flow. The Colloid Core is actually a liquid (or particulate) core concept.

It is the object of this thesis to review the latest Gas-Core Nuclear Rocket Designs, address the major design problems, and propose an optimum design based on current technology.

This paper is organized as follows; Section 1.0 is the introduction. Section 2.0 contains a review of previous Gas-Core Nuclear Rocket Designs. Problem areas and rocket performance parameters are outlined. Section 3.0 contains a review of the radiation problems associated with nuclear rockets. Section 4.0 contains details on a proposed method to improve nuclear fuel containment. Section 5.0 contains details on a proposed method to help maintain nuclear criticality. Section 6.0 outlines the proposed Gas-Core Design. Section 7.0 contains the summary plus recommendations for further study.

2.0   GAS-CORE NUCLEAR ROCKET TECHNOLOGY STATUS

The need for a Gas-Core Nuclear Rocket arises from the fact that chemical rockets are limited in their performance to specific impulses (I_{sp}) of about 500 seconds. And Solid-Core Nuclear Rockets are limited to an (I_{sp}) of about 1000 seconds. The latter is because the heat exchange/transfer process must take place thru a solid fuel containment barrier which is temperature limited to about $2500°$ K. However, the gas core nuclear rocket can achieve specific impulses up to 5000 seconds. This is because the heat exchange process takes place with the hydrogen propellant in direct contact with the uranium fuel (Colloid Core and Coaxial Flow designs) without the temperature limitations of the intervening solid wall.

2.1   COLLOID CORE DESIGN

The colloid fueled nuclear rocket engine utilizes fuels in the colloid form, ie. solid particles or liquid droplets with gaseous propellant. The suspension of the fuel is caused by the centrifugal force generated in the cavity. Intimate contact, and consequently good heat transfer, between the fuel and the propellant are maintained. This results in higher temperatures than those obtainable in the solid-core reactors. However, since significant vaporization would result in fuel losses, the propellant temperature is limited to a level lower than that of a gaseous-core reactor.

Four studies of Colloid Core Nuclear Rockets will be reviewed in this section. Results will be summarized in tables and figures for analysis and comparison with the results of subsequent designs.

"The concept was first evaluated by the Aerospace Research Laboratory (ARL) of Wright-Patterson Air Force Base. ARL developed a highly effective vortex chamber configuration which had an excellent separation capability. In conjunction with this development, an engineering study of the concept was also carried out." (REF 1). This study included performing a theoretical analysis of the reactor characteristics, evaluating the technological problems areas, defining a conceptual design of a ground test reactor (GTR), and recommending a test program for the reactor development. The reactor characteristics are given in table 1 below:

Thrust capability	100,000 lbf(45,000 kg)
Specific impulse(theoretical)	1200 sec
Static pressure at the inlet of exhaust nozzle	100-200 atm
Static temperature at the inlet of exhaust nozzle	3700°K-3900°K
Fuel Zr and U carbide alloy	(1U-10Zr)C
Cavity diameter	120 cm (47.5 in.)
Cavity fuel suspension zone length	18 cm (7 in.)
Radial reflector thickness	45 cm (17.7 in.)
Axial length of reflector	108 cm (42.5 in.)
Reflector annulus fueled zone thickness	15 cm (5.9 in.)
Pressure vessel outer diameter	240cm (95 in.)
Over-all length	240cm (95 in.)
Total engine weight (less storage)	19,000 kg (41,000 lb)

Table 1 - Characteristics and dimensions of the Colloid-Core Ground Test Reactor

3

Figure 1 shows the schematic diagram of the cavity reactor configuration and hydrogen flow paths (REF 1).

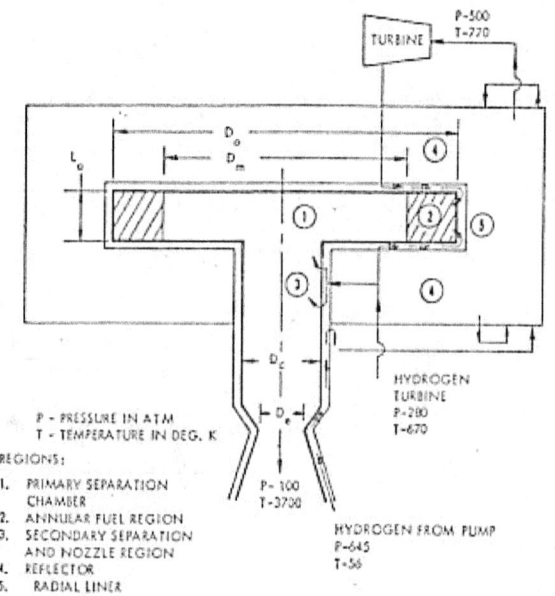

Figure 1 - Schematic of Cavity Reactor Configuration

The following description from (REF 1) outlines the conceptual design of a ground test reactor for this concept:

Fuel composition:

> "The use of alloy fuel particulates was based on a thermo-dynamic consideration. Because of the highly effective separation of fuel and propellant that can be achieved by a vortex chamber, the only loss of fuel considered was the uranium vapor carried out by the propellant. To minimize the vapor loss the fuel should have a low uranium equilibrium vapor pressure. The vapor pressure data of the ternary system involving U-C-Zr are based on the thermo-dynamic formulation of free energy expressed in terms of partial and vapor pressures of each element. The vapor composition in equilibrium with the solid fuel can then be computed at a given pressure and temperature. This yields the molecular weight of the vapor mixture from which the specific impulse of the rocket exhaust and the uranium loss rate can be evaluated."

Figure 2 shows the performance vs. pressure and temperature conditions. The solid lines represent loci of constant specific impulse and the dotted lines represent constant uranium loss rates. These loss rates are based on calculated vapor pressures that require experimental verification. "Within the range of specified pressure and temperature, the uranium loss rate is 28-30 Kg/min, corresponding to a uranium to hydrogen flow rate of 1/80." (REF 1).

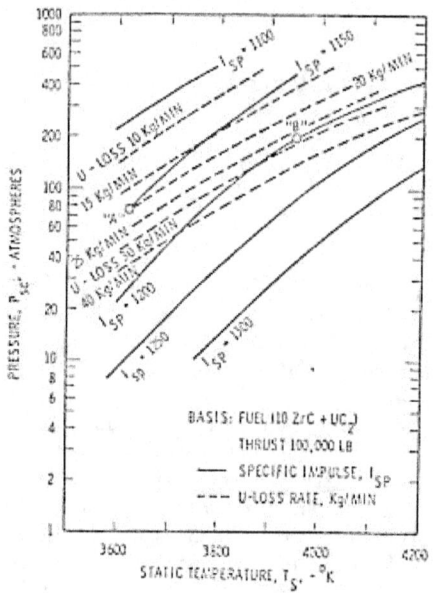

Figure 2 - Performance vs. Pressure-Temperature Conditions

U-233 vs. U-235 Fuel:

"The advantage of U-233 fuel as opposed to U-235 fuel can be determined by examining the microscopic cross-section data of both uranium isotopes. The average numbers of neutrons per fission of U-233 is slightly higher than that produced per fission of U-235, over the entire neutron energy spectrum range." (REF 1).

6

Particle Size and Fission:

"The slowing down of fission fragments in fuel material can cause ejection of fuel material from the fuel body. The fuel loss would be proportional to the number of fission fragments escaping from the fuel, and the latter decreases with increasing particle size. Assuming the ejected material is in the form of clumps containing 10^4 atoms of uranium on the average, and every tenth fission fragment causes fuel ejection, it was estimated for fuel particles of 100μ in diameter, at a flux of $10^{13}cm^{-2}sec^{-1}$ that the loss rate may be 10% of the core loading per minute. On the other hand, if the fuel particles are sufficiently reduced it is conceivable that the fission fragments may escape from the fuel particle before enough energy to displace 10^4 atoms has been transferred to the fuel. Also, a lower limit of size was based on fabrication restraints. Pending further development in the effects of the agglomeration and the fragmentation phenomena, a preliminary selection of the particle size in the range of 0.5-5μ was made for the GTR." (REF 1).

Vortex chamber:

"The vortex chamber or cavity configuration evolved from the hydro-dynamic experiments that were conducted by ARL. Figure 1(page 4) illustrates the chamber geometry thus defined and identifies major regions. The cavity region 2, having a length to diameter ratio (L_0/D_0) much less than unity. This characterizes this geometry as a compressed cavity. It has been shown to provide an increased capability for the retention of relatively large masses of solid particles. The primary function of the vortex chamber is to separate and retain fuel particles in the chamber with the solid-free gas exhaust at the exit end (D_e). Thus section (3) serves as a final clean-out stage of the exhaust gas. The finest particles in the exhaust gas stream can be centrifuged to the wall in region (3), where they will be carried to the closed end of the chamber." (REF 1).

This method of fuel containment will be discussed in more detail in section 4.0. The concept of extracting fuel from the fuel-propellant mixture just prior to entry into the exhaust nozzle area will be applied to the Coaxial-Flow Nuclear Rocket as well.

"Another consideration of the cavity configuration is the tendency of the particles to impinge on the peripheral wall. Because the centrifugal force acting on the particles varies with (V^2/R), increasing the radius R, with accompanying effect of a reduction in the tangential velocity V, will reduce the centrifugal force rapidly. Consequently, the tendency of particle impingement on the wall can be reduced by means of a large diameter D_0." (REF 1).

"Referring to Figure 1(page 4), assuming a uniform distribution of fuel in region (2), and a 4-mil tungsten liner placed in region (5). The axial and radial reflections, regions (4) and (5), are assumed 45 cm thick and made of beryllium. Of principle interest were the critical mass requirement and the resulting solid loading factors in the cavity fuel in region (2). The two-dimensional discrete ordinates transport theory code (DOT) was used in the nuclear calculations. Based on these, a configuration with $D_0 = 120$ cm and $L_0/D_0 = 0.15$ was chosen." (REF 1).

Conclusions reached from this study (REF 1):

1. Specific Impulse = 1200 secs.

2. Weight = 41,000 lb.

3. Thrust = 100,000 lb.

Another study performed by Anderson, Hasinger, and Turman (REF 2) quotes results of experiments with highly loaded solid-gas vortex flows. The apparatus they used is described as follows:

The vortex chamber used in the experiments is shown in Figure 3 below:

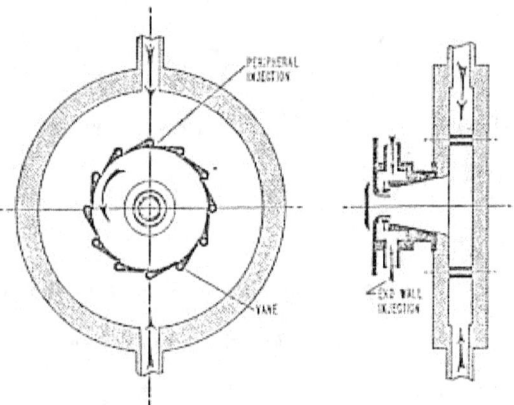

Figure 3 - End View and Side View of Vortex Chamber.

"It was made principally of Plexiglas to allow optical and x-ray observations of the flow. Air was injected tangentially at the periphery through 12 slits of 0.03 cm width, formed by 12 overlapping vanes extending the length of the chamber. The diameter of the chamber was 30.5 cm, with length at the periphery of 6.3 cm, giving a length to diameter ratio of 0.2. Talc particles were used for most of the experiments, with an average particle size of 20μ. The Talc density was $2.7 \frac{gm}{cm^3}$." (REF 2).

Powder loading and Containment Within the Vortex:

"Under normal operating procedures, vortex flow in the chamber was initiated with clean air. Then a separate vortex device was used to fluidize powder and pump it into the chamber through a hole in the end wall near the periphery." (REF 2).

Figure 4 summarizes observations of the functional relationship between maximum load and gas flow rate into the vortex. "As a reference value the maximum load attained here represents a particle volume fraction greater than 10%, with solid to gas mass density ratio in excess of 100:1." (REF 2).

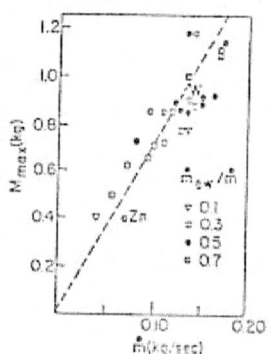

Figure 4 - Maximum Powder Load vs. Total Gas Flow Rate

"Powder loss rates were monitored within the chamber. The powder load was found to follow an exponential trend, so that a loss rate coefficient λ could be defined by:

$$M = M_0 e^{-\lambda t}$$

12

Where M_0 is the initial mass. An investigation into the effect of gas flow rates upon loss coefficient is summarized in Figure 5. The importance of end wall flow is dramatically presented in the graph (with $\dot{m}_{EW}$, where EW stands for End Wall."

(REF 2).

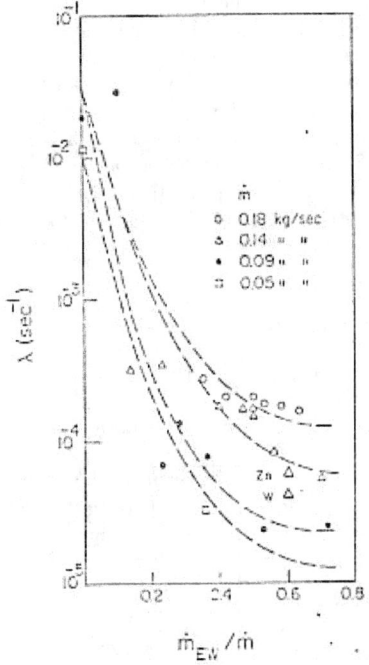

Figure 5 - Particle Loss Rate Coefficients vs. End Wall Fraction

13

Implications for the Colloid Core Reactor:

"As an example of a possible extrapolation from the present experiment
to operating conditions of the colloid core reactor, a hypothetical engine
will be assessed, with thrust of 20,000 lb force, and specific Impulse
of 1000 secs. A mass flow rate of 10 kg/sec would then be required.
Taking hydrogen as operating medium and assuming a temperature of
$3300° K$, the pressure within the cavity would be about 100 atmospheres.
A uranium carbide alloy, (1U-10Z)C, has been proposed as a possible fuel
for the colloid core reactor because of its low uranium equilibrium vapor
pressure. Material density of this fuel would be about $8.2\frac{m}{cm^3}$, near the
density of zinc. For this example the fuel particle diameter will be taken
as 10μ, the reactor radius is 30 cm, with an axial length of 30 cm, and
a inlet velocity of $100\frac{M}{Sec}$ is given for tangential gas injection. Critical
mass calculations for the cavity give a figure of 3 kg. uranium mass, for
a total fuel load of 30 kg. (REF 2).

The various extrapolated flow parameters for the reactor are compared with experimental conditions in Table 2 below. Note that the particle volume fraction required
by this fuel load is well below the 10% value attained in the given experiments.

Symbol	Experimental Vortex	Projected Reactor
Outer Radius of Vortex (R)	0.15 m	0.30 m
Height (h)	0.06 m	0.30 m
Gas Flow Rate per Unit Area ($\frac{\dot{m}}{A}$)	$1.7\frac{kg}{m^2 sec}$	$1.7\frac{kg}{m^2 sec}$
Gas Density (ρ_g)	$1.2\frac{kg}{m^3}$	$1.1\frac{kg}{m^3}$
Gas Viscosity (μ)	0.02 centipoise	0.04 centipoise
Propellant Density (ρ_p)	$7.0\frac{gm}{cm^3}$	$8.0\frac{gm}{cm^3}$
Particle Diameter (D)	$10\ \mu$	$10\ \mu$
Total Power Mass (M)	0.8 Kg	30 kg
Fraction(1 − ϵ) ϵ	0.9	0.95
Radius (r)	0.08 m	0.13 m
Angular Velocity (ω)	$150\frac{rad}{sec}$	$450\frac{rad}{sec}$

Table 2 - Flow Parameters

14

"The power output required by the engine performance specification is about 600 MW, with a reactor power density of about $10^4 \frac{MW}{M^3}$. A fuel loss rate of about $30 \frac{gm}{sec}$ is predicted, for a uranium loss rate of $3 \frac{gm}{sec}$. Vaporization loss for uranium has been estimated at roughly $100 \frac{gm}{sec}$ at an operating temperature of $3000°K$, so the particulate loss will be small in comparison to vaporization." (REF 2).

Other experiments with heavily loaded two-component vortex flows show that the interpretation of particle behavior for small length/diameter ratios, is valid for length/diameter ratios approaching unity. The rotating fluidized bed model allows extrapolation of these experiments to the operating regime for a conceptual colloid core reactor.

Some projected values for the operating conditions of such a reactor (REF 3) are listed in Table 3 below:

Thrust	22,000 lbf
Specific impulse	$1,000 \frac{lbf sec}{lbm}$
Chamber radius	$R = 30cm$
Chamber length	$h = 30cm$
Fuel mass (1U-10Zr)C	$M = 20kg$
Gas flow rate	$m' = 10\frac{kg}{sec}$
Gas injection velocity	$v_0 = 50\frac{m}{sec}$
Fuel volume fraction	$\nu = 0.03$
Inner bed radius	$r = 13cm$
Inner bed velocity	$\omega = 300\frac{rad}{sec}$
Fluidizing pressure drop	10 atm

Table 3 - Projected Reactor Parameters

Table 4 contains performance potential calculations for a Colloid Core rocket engine when applied to the upper stage of the Orbit-to-Orbit Shuttle Space Transportation System (REF 4). These are compared to the performance potential of a representative Solid Core Rocket OOS Vehicle. Typical OOS missions are between 100 naut-mile parking and Geosynchronous orbit.

Parameter	Small Solid Core Reactor	Nominal Performance CCNR
Specific impulse	860 Sec	1100 Sec
Thrust	16,135 lbs	20,000 lbs
Fuel loss rate	$0 \frac{lbm}{sec}$	$0.147 \frac{lbm}{sec}$
Initial propellant	30,441 lbs	34,500 lbs
Tankage fraction	0.052	0.052
Engine weight	5103 lbs	3311 lbs
Shield diameter	2.116 ft	1.378 ft
Shield constant	$118 \frac{lbm}{ft^2}$	$200 \frac{lbm}{ft^2}$
Structural weight	2596 lbs	2596 lbs
Start up/shutdown	33 Sec	2.77 Sec

Table 4 - Comparative O.O.S. Vehicle Characteristics

Summary:

In summary the Colloid Core Nuclear Rocket seems to be a reasonable design. The projected reactor size is modest, and there appears to be no major problems associated with rotating the required fuel load. The reactor power density is well within the heat transfer capacities of modern cooling systems. The uranium loss rate seems to be the major area of concern and this will be further addressed in Sections 4 and 6.

2.2   NUCLEAR LIGHT-BULB DESIGN

The next nuclear rocket design considered is the Nuclear Light Bulb Rocket; a true gas core design for the uranium fuel is completely in a gaseous state. In the nuclear light-bulb engine, the fissioning plasma is confined within a transparent cell and kept away from the walls of that cell by a swirl-flow of tangentially injected buffer gas. The transfer of power from the fissioning fuel to the propellant is by radiative heat transfer. This is limited by the transparency of the confining quartz structure which restricts radiative energy fluxes to a certain wavelength region. However, this concept offers complete fuel containment, see Figure 6 below.

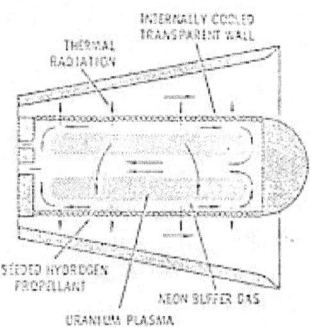

Figure 6 Nuclear Light Bulb Chamber

This is one reaction chamber. The complete Nuclear Light Bulb Engine consists of seven modules such as shown in Figure 7.

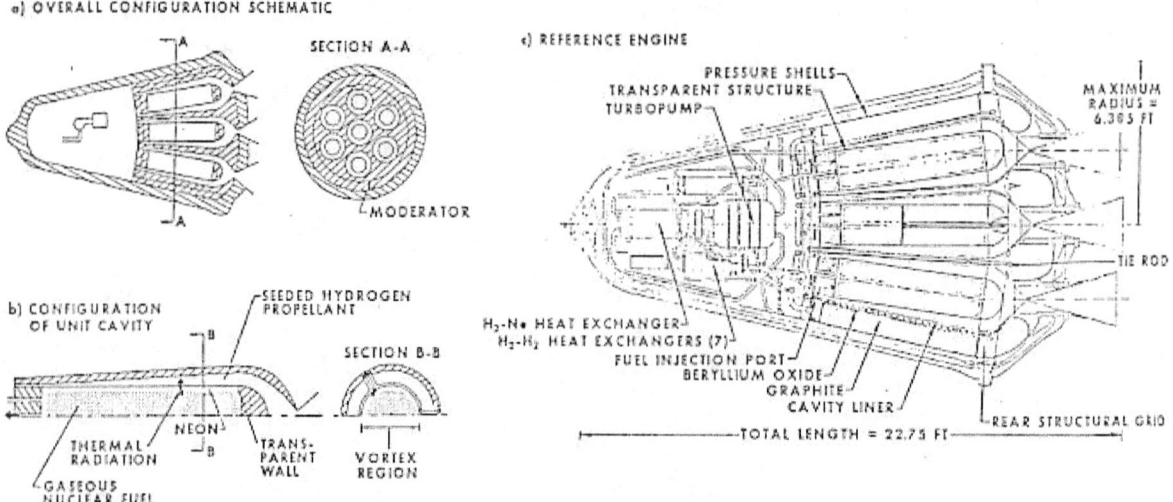

Figure 7 Nuclear Light Bulb Engine

Two studies were reviewed and the results are summarized below. In a review of Gas Core Nuclear Rockets (REF 5), McLafferty describes in detail a Nuclear Light Bulb Engine design;

> "The gaseous nuclear fuel is isolated from the transparent wall by a neon vortex. This neon flow passes out through ports located on the centerline of the end wall of each cavity to a fuel recycle system, wherein nuclear fuel entrained in the neon is condensed to liquid form, centrifugally separated from the neon, and pumped back into the fuel region.

> In the reference engine, each of the seven cavities has a length of 6 feet. The total volume of all seven cavities is equal to that of a single cylinder having a diameter of 6 feet and a length of 6 feet. The total amount of fuel contained within the seven cavities is ~14 Kg, and the power is ~ 4600 MW. The total pressure in the cavity is estimated to be 500 atmospheres. The total hydrogen flow rate of $49 \frac{lb}{sec}$ is heated to $12,000°R$, which will provide a specific impulse of 1870 seconds. The resulting engine thrust is therefore, 92,000 lb. The total weight is estimated to be 70,000 lb and is made up of the following component weights: moderator (graphite and beryllium oxide), 27,000 lbs; pressure vessel, 30,000 lbs; turbopumps, 3000 lbs; and miscellaneous (including the fuel recycle system), 10,000 lbs."

One of the primary objectives of the fluid mechanics work conducted under the nuclear light bulb program was to provide a flow geometry which will prevent condensation of the gaseous nuclear fuel on the transparent wall. Results obtained in unheated gas vortex tests quoted in REF 5 are presented below.

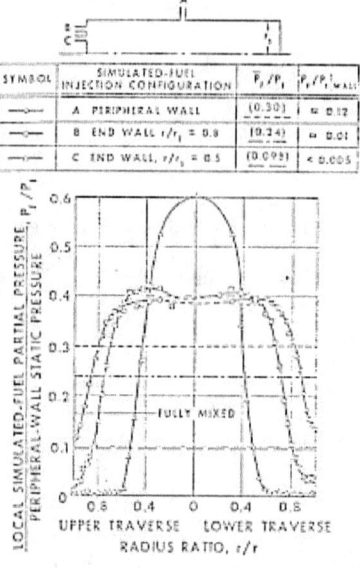

Figure 8 Radial Distributions of Simulated Fuel

"The results shown indicate that the location of the simulated fuel injection has a major effect on the partial pressure distribution of the simulated fuel in the vortex tube. Moving the fuel injection port radially inward on the end wall results in an increase in simulated-fuel partial pressure near the centerline and a decrease in simulated-fuel partial pressure near the outer periphery; however, this also results in a decrease in the average simulated-fuel partial pressure within the vortex tube. It would appear from Figure 8 that a radius ratio somewhere between 0.5 and 0.8 will result in a better combination of high average simulated-fuel partial pressure and low partial pressure of simulated-fuel near the peripheral wall than any of the three injection configurations shown in Figure 8. Tests have also been conducted with RF heated vortexes in which the radial temperature gradient near the outer periphery results in a further decrease in simulated-fuel partial pressure near the peripheral wall." (REF 5).

The RF induction heating equipment at the United Aircraft Research Laboratories has been used to obtain a very intense radiation source for use in demonstrating that transparent wall structures can be successfully cooled, and for use in tests in which a simulated propellant is heated by thermal radiation. Results from some of these tests are given in Figure 9 below. (REF 5).

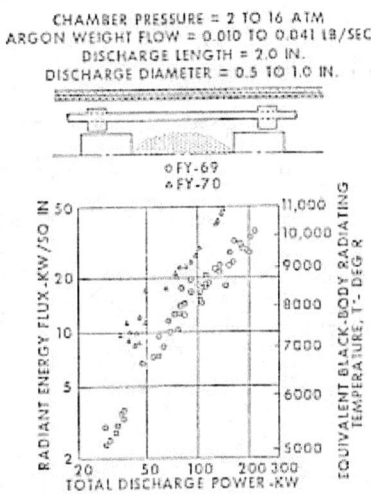

Figure 9 Tests of RF Radiant Energy Source for Nuclear Light Bulb Investigations

"During tests conducted in 1969 a peak radiant flux at the edge of the RF discharge of $36 \frac{kW}{in^2}$ was obtained from an ellipsoidal discharge region which was 2 in. long and which had a diameter of ~0.8 in.(corresponding radiated power of 156 kW and total deposited power of 216 kW). This radiant heat flux is equal to that from a blackbody at a temperature of $10,200°R$ ($5650°K$). Preliminary tests conducted in 1970 resulted in a reduction in the diameter of the discharge, and a resulting increase in flux for a given power level. The peak flux obtained to date (Dec 1970) from these preliminary tests is $48 \frac{kW}{in^2}$ (see Figure 9) which corresponds to a blackbody radiating temperature of $10,800°R$. This flux is slightly greater than the flux at the surface of the sun. The energy radiated from the plasma in this test was 85% of the energy deposited in the plasma." (REF 5).

One of the objectives in generating the high intensity radiant energy source shown in Figure 9, was to permit tests to be conducted in which seeded simulated propellant is heated by radiation from the high intensity source. "The highest indicated bulk exit temperature obtained in tests to date (Dec 1970) was $2200°R$. A comparison of results of fluid mechanics research with the requirements of full-scale nuclear light bulb engine is given in Table 5 below". (REF 5).

Research Area	Simulation parameter	Value for full-scale engine	Level achieved to date in research program
Propellant Heating by thermal radiation	Exit Temp. $^\circ$R	12,000	2,200
Transparent-wall models	Wall thickness, in.	0.005	0.005
	Heat deposition rate, $\frac{kw}{in.^3}$	1.6	2.2
Radiant energy source	Radiant flux, $\frac{kw}{in.^2}$	178	47.9
Two-component vortex tests	Simulated-fuel partial pressure fraction	0.25	0.1(rf tests) 0.4(constant temperature tests)

Table 5 Comparison of results of fluid mechanics research with the requirements of full-scale nuclear light bulb engine.

Radiant Heat Transfer:

"On the subject of radiant heat transfer, the opacity of the gaseous nuclear fuel is so high that the centerline temperature is much greater than the blackbody radiating temperature $\sim 60,000°R$." (REF 5)

McLafferty also addresses the use of seeded material to enhance radiation absorption at low temperatures. One possible material is tungsten in the form of small-diameter particles whose total mass is several percent of the mass of the hydrogen propellant.

The transparent wall dividing the fuel and propellant regions must allow the transmission of energy without too much absorption. REF 5 states that this energy is mostly contained in the wavelength region between .1 and 4.0 μ.

2.2.1   NUCLEAR INDUCED COLORATION

Another possible problem is nuclear induced coloration of the transparent wall. Some measurements of the transmissions characteristics of fused silica (REF 5) indicate that exposure of the specimen to neutron and gamma radiation resulted in an increase in the absorption coefficient in the ultraviolet region. However, heating of the specimen to $\sim 800°C$ resulted in annealing of this radiation induced coloration back to its pre-radiation value. If the absorption spectrum of the transparent wall at $800°C$ is used with the blackbody radiation spectrum for a temperature of $15,000°R$ (the radiating temperature in the representative nuclear light bulb engine) than it is calculated that 1% of the incident energy is absorbed.

The problem with such investigations is that it is extremely difficult to simulate the neutron and gamma ray flux which is present in a full-scale engine.

In summary it appears that the most critical question is that of the radiative energy transfer through the confining transparent wall. The limited transparency of the quartz structure restricts radiative energy fluxes to a certain wavelength region. On the other hand, the concept of the Nuclear Light Bulb Engine offers complete fuel containment; ecological restrictions may make this design the vehicle of choice for near Earth missions.

25

2.3   NUCLEAR COAXIAL FLOW DESIGN

In this concept, like the nuclear light bulb, the nuclear fuel is in a gaseous state. However, there is no transparent wall separating the nuclear fuel from the hydrogen propellant. Nuclear fuel is maintained within the cavity by purely fluid mechanical means.

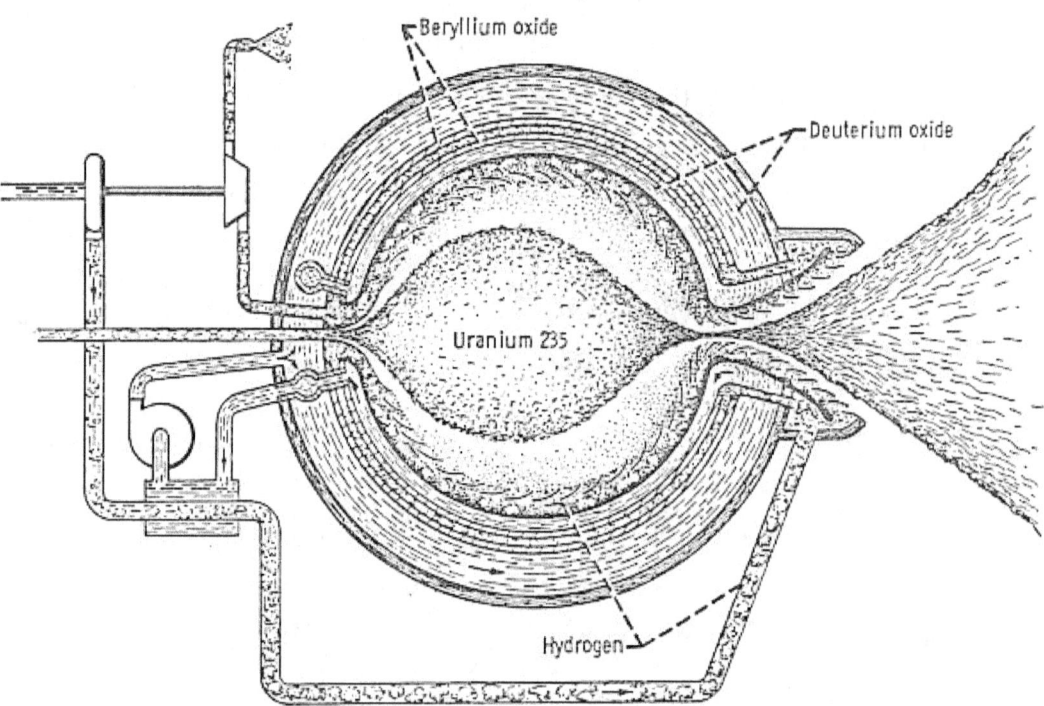

Figure 10 Coaxial Flow Gas Core Nuclear Rocket

As Mclafferty describes it in REF 5;

"In this concept a low-velocity cloud of gaseous fissionable fuel is surrounded by a higher-velocity stream of seeded hydrogen propellant. The seeded propellant is heated by thermal radiation from the hot gaseous nuclear fuel. The cavity containing the fuel and propellant is surrounded by a moderator region to reflect the neutrons created by the fission process back into the cavity to sustain the nuclear chain reaction. The hot hydrogen propellant, along with a small quantity of unburned nuclear fuel and fission products, is exhausted through a transpiration-cooled nozzle to provide thrust. Extensive analysis has been conducted to optimize engine size, engine pressure, etc.

A representative configuration might employ a spherical chamber having a diameter of 12 feet and containing 50 Kg of Uranium 233. The total power created in the nuclear reaction, 22,000 MW, would be sufficient to heat $222 \frac{lb}{sec}$ of Hydrogen to a temperature which would result in a specific impulse of 1800 sec., thus providing a thrust of 400,000 lb. The required pressure within the cavity would be ~ 1000 atmospheres, and the fuel loss rate would be $\sim 1\%$ of the hydrogen flow rate. The moderator surrounding the cavity would be composed of heavy water and beryllium oxide, with a wall thickness of 2.5 feet. The total weight of this representative configuration is estimated to be 281,000 lb, which is made up of the following component weights: moderator, 120,000 lb; pressure shell, 140,000 lb; turbopump, 19,000 lb; and exhaust nozzle, 2000 lb. The centerline temperature for the Coaxial Flow Reactor is $\sim 100,000°$ R."

In REF 6 F. Rom addresses the nuclear criticality problem as it relates to nuclear fuel containment;

> "Criticality determines the amount of fuel that is required in the reactor to maintain the chain reaction. Inasmuch as the fuel is in a gaseous state, the criticality requirement, which prescribes the number density of fuel atoms, is a determining factor of the pressure in the reactor.
>
> The fluid mechanics problem area is concerned with minimizing the fuel loss rate. In the gas core, if the uranium and hydrogen are completely and thoroughly mixed, the uranium loss rate would be prohibitive. Fluid mechanics studies are required to determine ways and means for increasing the residence time of the uranium while the hydrogen flows through the reactor as quickly as possible (while entraining a minimum amount of uranium). Producing and understanding flow fields that do this is the major challenge to a successful Gas Core design."

The Criticality Experiment Test Installation, Lewis Research Center (performed at the National Reactor Testing Site of the Atomic Energy Commission in Idaho by the General Electric Company), was setup to test the feasibility of a gas core design. From REF 6; the experiment is described as consisting of a deuterium oxide reflector moderated cavity. The cavity was 6 feet in diameter and 4 feet long. The deuterium oxide reflector was 3 feet thick. The outside diameter of the reactor was therefore 12 feet, and the length was 10 feet. Approximately 20 tons (18,100 Kg) of deuterium oxide is required in this cavity. Results of the experiment are shown in Figure 6 below.

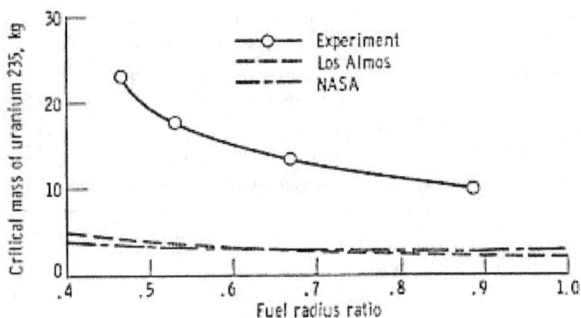

Figure 11 Cavity Critical Experiment Results

The results of these tests (from REF 6) are as follows;

"The critical mass in kilograms of uranium 235 is plotted as a function of fuel-to-cavity radius ratio. The experimental points are indicated by circles with the solid line drawn through them. The estimates made previously by NASA and Los Almos are shown with dashed lines. The previous estimates show critical masses in the range of 3.5 to 4.0 kilograms for a fuel-to-cavity radius ratio from 1 down to 0.5. There exists a large difference between the calculated and the experimental values of the critical mass. The critical masses are from three to four times the calculated values. This means that the pressure levels within the reactor will be three to four times what was previously predicted.

Extrapolations based upon the previous experiment are as follows: Figure 12 shows some cavity reactor characteristics as affected in going from a room-temperature critical experiment to a hot operating reactor. The case we have considered is a deuterium oxide reflected cavity 8 feet in diameter and 8 feet long. The fuel region is 6 feet in diameter, while the reflector thickness is 3 feet, giving an outer diameter of 14 feet. The first case shown assumes that there is no hydrogen in the core and that the uranium is vapor at room temperature. The critical mass of the uranium would be 9 kilograms. Uranium in a gaseous state at room temperature would yield a pressure of 0.19 atmospheres ($19,250\frac{N}{M^2}$). If we introduce hydrogen at a temperature of $530°$R ($294°$K). With an atom density of 10^{21} atoms per cubic centimeter in the region between the fuel and the cavity walls, while maintaining the uranium at room temperature, the critical mass would be increased by 11.2 kilograms because of the presence of the hydrogen. This increases the pressure to 0.24 atmospheres ($24,300\frac{N}{M^2}$).

The next case considers the effect of operating temperature levels. The average hydrogen temperature within a gas core with a specific impulse of 1500 seconds would be about $6000°$R, ($3330°$)K. The average uranium temperature would be about $90,000°$R, ($50,000°$K). The primary effect here is to increase the critical mass to 16.8 kilograms because of the upscattering of neutrons (to higher temperature) by the hot hydrogen.

The uranium pressure in now 54.4 atmospheres ($5,520,000\frac{N}{M^2}$) assuming that the uranium was not ionized. The next line indicates the increase in pressure due to the ionization of the uranium. At this temperature the uranium would be more than triply ionized. The pressure is then 209 atmospheres ($21,200,000\frac{N}{M^2}$). All these calculations were carried out assuming that there was no structural material between the deuterium oxide and the uranium. Actually, some structure would have to be provided to contain the deuterium oxide. And if it is assumed that this structure is equivalent to 1.5 centimeters of aluminum, the critical mass increases to 32.3 kilograms, which results in a pressure of 404 atmospheres ($41,000,000\frac{N}{M^2}$)."

TABLE III. - GAS-CORE OPERATING PRESSURE ESTIMATES

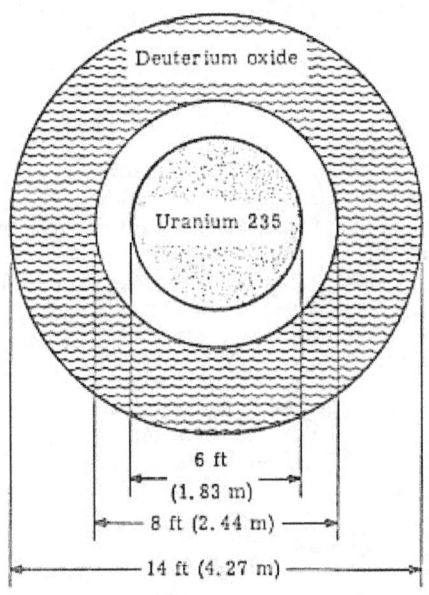

6 ft
(1.83 m)

8 ft (2.44 m)

14 ft (4.27 m)

| | Temperature | | | | Critical mass, kg | Minimum pressure | |
| | Hydrogen | | Uranium | | | atm | N/m^2 |
	°R	K	°R	K			
Fuel	None	----	530	295	9.0	0.19	19 300
	[a]530	295	530	295	11.2	.24	24 300
	6000	3330	80 000	44 400	16.8	54.4	5 520 000
	6000	3330	80 000	44 400	16.8	209	21 400 000
Structure	----	----	-----	-----	32.3	404	40 900 000
Fuel distribution	----	----	-----	-----	47.8	595	60 300 000
50 Percent containment	----	----	-----	-----	47.8	1190	120 600 000

[a]Hydrogen number density, 10^{21} atoms/cm^3.
[b]Assumes no ionization of uranium 235.

Figure 12 Gas Core Operating Pressure Estimates

"The fact that the fuel is not uniformly distributed within the fuel region must also be considered. Critical experiments at the National Reactor Testing Station in Idaho were run in which uranium was distributed as it might be in a typical gas-core reactor. The critical mass increased by 50% because of this effect. This would increase the pressure to 595 atmospheres ($60,300,000 \frac{N}{M^2}$).

Finally, in any gas-core system the uranium would not be completely separated from the hydrogen as shown in the schematic drawing. There would be a certain amount of hydrogen that would mix with the uranium. The uranium concentration could easily be reduced by 50% as a result of this mixing. If there is no change in critical mass, the pressure would be doubled, giving 1190 atmospheres ($120,600,000 \frac{N}{M^2}$)."

Gas-Core Fluid Mechanics:

The basic coaxial flow model is shown in Figure 13 below.

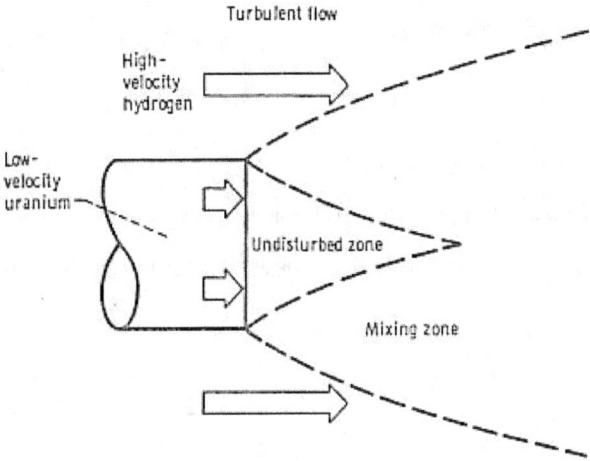

Figure 13 Basic Coaxial Flow Model

Low velocity uranium enters through the center duct, while high velocity hydrogen flows around this duct. Because of the velocity difference, there will be a mixing between the hydrogen and uranium. This results in what is called a mixing zone. There also be an undisturbed zone as shown. The undisturbed zone is formed by the inner boundaries of the mixing zone. The undisturbed zone is where most of the fuel would be contained. Several coaxial flow mixing experiments have been carried out both at Lewis and at the Illinois Institute of Technology (IIT) under contract to Lewis. In addition, theoretical calculations have been made to determine the

33

flow fields that exist in a coaxial-flow situation. Idealized coaxial-flow experiments have been run at Lewis and IIT. At Lewis experiments were run with bromine and air. "Bromine was injected in a low-velocity central region to represent the uranium while air was flowed around this region at high velocity to represent the hydrogen. The velocity ratios and flow rates were varied over a wide range to obtain data for an analytical correlation required for prediction of coaxial flows at other conditions." (REF 6).

"Similar data have been taken recently at IIT using freon as a heavy gas simulator. One of the results of this experiment is shown in figure 14 below." (REF 6).

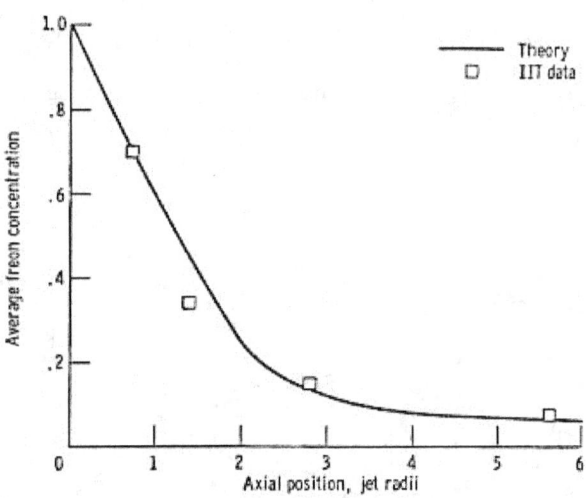

Figure 14 Freon Concentration vs. Axial Position

"Here, the average freon concentration is plotted as a function of axial downstream position, in units of jet radii. The experimental points are indicated by the square data points, for an initial velocity ratio of 31 to 1; that is; the air is flowing 31 times the velocity of the freon. The theoretical prediction, based on some work at Lewis, is shown superimposed on these data. The agreement is good." (REF 6).

"Figure 15 below shows the results of the theoretical calculations for the case of a velocity ratio of 30 to 1, and a mass flow ratio of 35 to 1, which is representative of a gas-core reactor. Shown are relative concentration profiles for 95%, 70%, 40%, and 10% fuel. Most of the fuel is concentrated in the undisturbed region near the entrance." (REF 6).

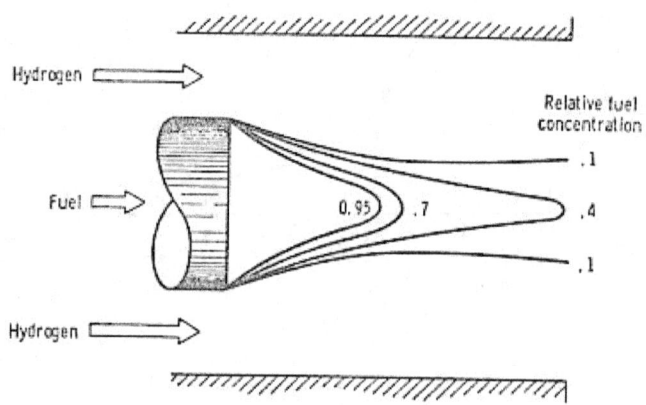

Figure 15 Concentration Field in Coaxial Flow Engine

The purpose of the above figure is to indicate that the concentration distributions can be calculated for isothermal conditions within the core with a theory that has been substantiated by laboratory experiments. It is also possible to calculate the corresponding velocity profiles throughout the core region.

The central problem of interest is the calculation of radiant heat transfer between the uranium fuel and the hydrogen propellant. The accurate determination of the absorption coefficient of the gases involved must be determined so that the heat transfer may be calculated.

From REF 6 Figure 16 (below) shows examples of the absorption coefficients as a function of temperature for the various regions in the reactor. In the nuclear fuel region the absorption coefficients are very high compared with the hydrogen propellant. In the case of the hydrogen propellant the absorption coefficient becomes very small at temperatures below $10,000°R$ ($5550°K$). At $40,000°R$ ($22,200°K$) it reaches a peak value and again starts to fall off in the $100,000°R$ ($55,500°K$) range. In the gas-core, the hydrogen enters at a relatively low temperature. In this case the hydrogen itself cannot absorb radiation; therefore, particulate (seed) must be introduced into the hydrogen to render it opaque to thermal radiation. A typical absorption coefficient for hydrogen seeded with particles is shown in Figure 16.

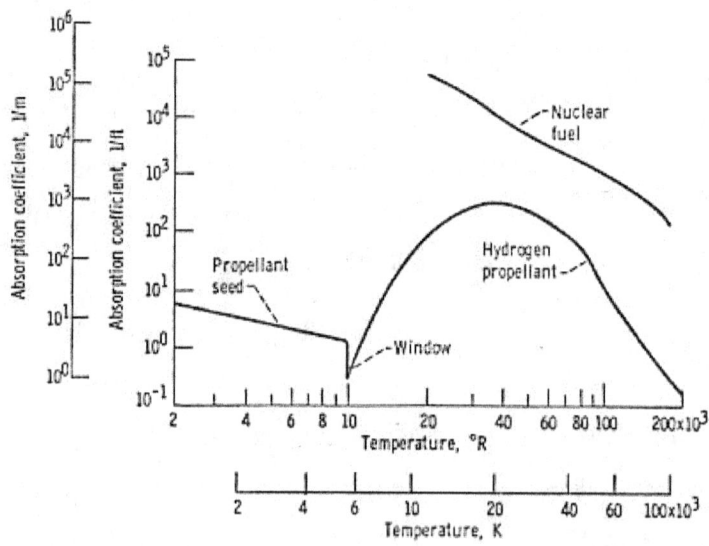

Figure 16 Absorption Coefficients vs. Temperature

When the propellant seed increases in temperature toward 10,000°R (5500°K) it will evaporate and may leave a film if the vapor of the seed material does not absorb radiation.

Figure 17 shows the calculated results of the temperature field in a coaxial-flow engine. Considered is both the non-uniform fuel distribution and non-uniform heat generation that would exist in a real case. Shown here are the isotherms for $4,000°R$, $7,000°R, 9,000°R, 11,000°R, 80,000°R, 100,000°R$, and $120,000°R$, $(2,200°K, 3,890°K, 5,000°K, 6,110°K, 44,400°K, 55,000°K$, and $66,600°K)$. "The incoming hydrogen temperature for this case was $3,500°R$ $(1,945°K)$. Seeding was introduced into the hydrogen to render it opaque in a low-temperature region." (REF 6)

Figure 17 Temperature Field in Coaxial Flow Engine

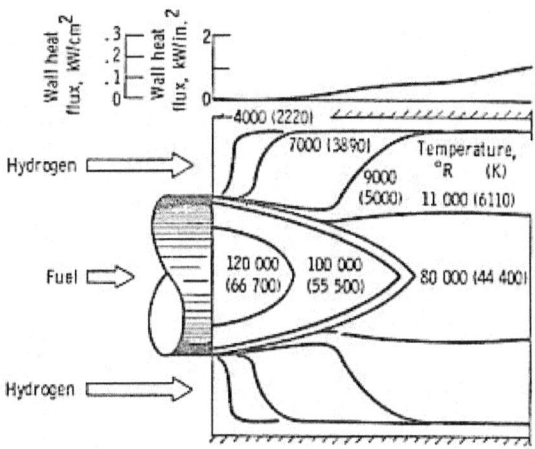

"The isotherms bend toward the wall as the temperatures get into the 7000°R, 9000°R, and 11,000°R (3890°K, 5,000°K, and 6,110°K) range because of the evaporization of the seed at 11,000°R (6110°K) the hydrogen becomes opaque, and the contours once again come toward the center. The maximum fuel temperatures are in the range of 120,000°R (66,600°K) and the average fuel temperatures are about 90,000°R (50,000°K). The average hydrogen temperature will be about 10,000°R to 20,000°R (5550°K to 11,100°K) to produce specific impulses in the range of 1500 to 2000 seconds. The amount of seed material required to produce this temperature profile was less than 1% of the hydrogen mass flow, introducing a negligible effect on the specific impulse. This seed also was sufficient to reduce the heat flux to the wall to less than 1 kilowatt per square inch ($0.155 \frac{kW}{cm^2}$), which can be readily handled by a cooling system. This calculation indicates that as far as heat transfer problems are concerned, the coaxial-flow reactor is feasible." (REF 6).

Rough estimates of the performance that might be obtained from coaxial-flow gas-core reactor are shown in Table 6 below:

Parameter	
Thrust	250,000 lbf, 1,112,000 N
Weight	100,000 lbm, 45,360 kg
Thrust per Engine Weight	2 - 20
Specific Impulse	1500 to 2500 Sec
Pressure Level	1000 atm, 101,330,000 $\frac{N}{m^2}$
Uranium-to-hydrogen flow ratio	0.01 to .10
Average Uranium Temperature	$\sim 80,000°R$, $\sim 44,400°K$

Table 6 - Gas Core Performance

"These estimates are based on calculations using theoretical techniques that have been checked by experimental studies. The thrust-to-engine weight ratio will vary from 2 to 20. There is very little increase in fuel temperature required to increase the thrust by a factor of 10. (The thrust increase is obtained by increasing the hydrogen flow.) Because the power is radiated to the propellant, the fuel temperature increases approximately as the square root of the required reactor power. The hydrogen outlet temperature will be in the range of $10,000°R$ to $20,000°R$ $(5550°K$ to $11,100°K)$." (REF 6).

2.3.1 COAXIAL FLOW BUOYANCY EFFECT:

Another problem with the fluid mechanics involved in the Coaxial Flow Design is the Buoyancy Effect. In this analysis the inertial mass of the fuel region must be accelerated with the rocket and remain in the center of the reaction chamber with a spherical shape, in order to maintain nuclear criticality, fuel containment, and correct heat transfer characteristics to the hydrogen propellant. One study performed by REF 7 has the following results:

"In a typical coaxial-flow engine the low fuel velocities of about 0.01 f.p.s., coupled with a vehicle acceleration of about $g = 0.02 g_0$ (g_0 is the earth's gravity constant), could cause a buoyancy effect which could also reduce the fuel containment in the cavity."

"The cavity flow was analyzed using a Navier-Stokes equation numerical solution. The purpose of the analysis was to determine a scaling parameter for the buoyancy effect so that experiment conditions at 1.0 g_0 can be related to the actual engine flow conditions. The flow parameters for an engine, for experiments, and for this analysis are given in Table 7 below. The parameter B is the buoyancy scaling parameter." (REF 7).

	Engine Stage	Coaxial Flow Experiment	This Analysis
$\frac{m_r}{m_f}$	> 50	30-300	50
$\frac{\rho_t}{\rho_r}$	5-15	1.0,4.7	1.05,2.0,4.0,10
R_e	$10^4 - 10^6$	2×10^5	1000
B	50-500	2-30	0-100
VF	> 0.20	< 0.34	≤ 0.072

Table 7 - Range of Flow Conditions

"The cavity flow model to be analyzed is shown in Figure 18. The analysis is for a rectangular cylinder geometry with a cavity length-to-diameter ratio, ($\frac{L}{D}$), of one. This is similar to the cavity geometry of the coaxial flow experiments. The assumptions of the analysis are steady, isothermal, two-fluid, viscous flow with purely axial inlet and outlet flows and with no-slip walls. The system of equations to be solved includes the variable density Navier-Stokes equations including acceleration body force ρg, the continuity equation, and the mass transfer equation." (REF 7).

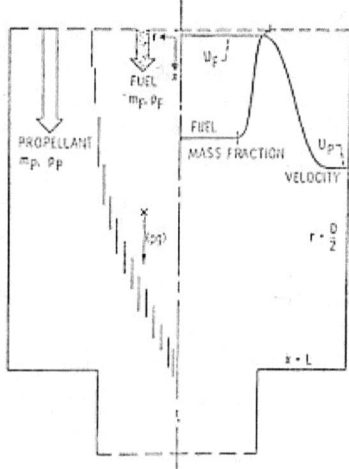

Figure 18 Cavity Model

For inlet conditions the inlet velocity and mass fraction profiles were specified as smooth profiles with buffer layers (See Figure 18 above). The total fuel mass in the cavity is obtained by volume integration of local fuel density, y_ρ, and the fuel volume fraction is obtained from:

VF = total fuel mass in cavity / pure fuel density X cavity volume

"The general effect of the vehicle acceleration on the fuel containment is illustrated by the calculated fuel mass fraction contours in Figure 19 for a density ratio of 2.0. Figure 19a is for zero acceleration, or $B = 0$, and Figure 19b is for an acceleration in the range of engine values with $B = 70$. This shows that the fuel region is stretched out and becomes narrower as the acceleration increases. the net result is a decreased fuel volume fraction and a significant buoyancy effect." (REF 7).

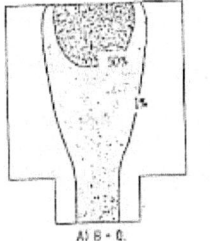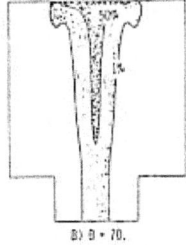

Figure 19 Fuel Mass Fraction Contours for Different Vehicle Accelerations.

Effect of acceleration on fuel volume fraction:

"Calculated fuel volume fractions for various accelerations over the range of density ratios are plotted in Figure 20 below. It was concluded on the basis of various cross-plots and Figure 20 that the best scaling parameter for generalizing the buoyancy effect has the form:

$$B = [(\rho_f - \rho_p)\frac{gD}{\rho_f U_f^2}]^{0.5}$$

This parameter will be called the buoyancy number. The combination inside the square root is a measure of the ratio of buoyancy force to inertia force. The square root is used for convenience in plotting the large range of B values." (REF 7).

"The calculated fuel volume fractions are thus plotted in Figure 20 as a function of buoyancy number. At a fixed density ratio the fuel volume fraction is constant for B values below about 5, and decreases as a constant power of B for large B values. Of particular importance is the fact that, for the specific combination of variables in B, the curves for various density ratios fall close together at large B values." (REF 7).

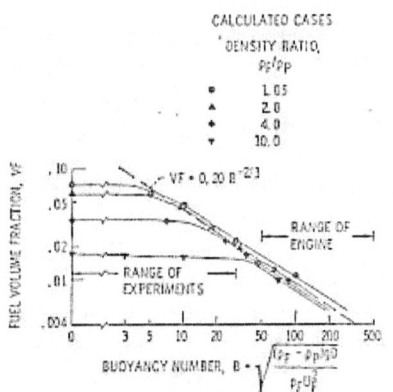

Figure 20 Fuel Volume Fraction vs. Buoyancy Number

"At buoyancy numbers above approximately 30, the fuel volume fractions for all density ratios are close to a single exponential line and may be represented by a signal correlating equation in the form:

$$VF(B >= 30) = 0.20B^{-\frac{2}{3}}.$$

This line is used to extrapolate the calculations to the buoyancy number of 350 for an "0.02 g" engine with a density ratio of 10 for this engine. As shown in Figure 20 the analysis predicts that the buoyancy effect decreases the fuel volume fraction by a factor of about 4. This calculated factor is subject to the limitation of the analysis. However, it indicates that a strong buoyancy effect may occur." (REF 7).

47

2.3.2   ACOUSTIC INSTABILITIES

Acoustic instabilities which have occurred in chemical rocket combustion chambers may become a problem in the gaseous core nuclear rocket as well. The purpose of the investigation carried out in REF 8 was to determine if such acoustic instabilities could indeed occur in a gas core nuclear rocket and to determine appropriate stability criteria.

> "Complex dynamic systems can be described by a set of field variables which can express the essential characteristics of the system in terms of the interaction of representative fields. The essential characteristics of the fields can be expressed as a set of non-linear partial differential equations derived from the conservation laws of mass, momentum, energy, neutronics, and radiation. Small signal perturbation theory can be used to investigate stability. A solution is obtained by initiating a disturbance and solving the set of equations." (REF 8).

In REF 8 a relationship between the critical wavelength for instability and the critical size for instability of a finite-length reactor cavity was developed. "The relationship between wavelength and size depends on the boundary conditions appropriate for the finite cavity. The nature of the boundary conditions is determined by the presence and specific characteristics of nozzles, injection systems, etc.

The data in REF 8 is based upon the assumption, that the fundamental standing wave will have a wavelength twice the length of the cavity (ie. the size of the cavity is the distance between nodes of the wave).

"To determine the practical implications of the stability criterion, a system with typical characteristics was analyzed. The critical wavelength was found to be about about 100 cm, so instability is expected in a system more than 50 cm in length.

However, a typical value of a core length is 300 cm. Thus, a typical reference design would be unstable with respect to acoustic waves." (REF 8).

In conclusion, acoustic instability is a potential problem for gaseous core nuclear rockets and merits further study.

Summary:

Because the power is radiated directly to the propellant without any intervening structure, the Coaxial Flow Design would seem to make the best use of the plentiful energy available in the nuclear reaction. The major area of concern here is the fuel loss rate or the Fuel-to-Propellant mass ratio. This will be further addressed in Sections 4 and 6.

3.0 RADIATION EMITTED

As the plume from the coaxial-flow gas core nuclear rocket forms, the crew is exposed to gamma radiation from the fission fragments in the plume. REF 9 estimated the radiation dose rate and the total dose to the crew from fission fragments in the plume for four specific missions to the planet Mars.

Other sources of radiation include direct radiation from the reactor core. This, plus solar radiation must be added to the plume radiation to obtain the total crew dose.

The four trip times selected in REF 9 were 80, 100, 150, and 200 days. The mission selected was the so-called "courier" mission to mars. This trip involves leaving a 600 km circular orbit around Earth with a trans-Mars injection maneuver, a high ellipse orbit insertion maneuver at the planet Mars, a Mars orbit time of approximately one day, a trans-Earth injection maneuver, and a circular orbit insertion maneuver upon return to Earth. The "courier" mission implies no payload will be sent to the planet's surface.

For each mission a specific engine with it's own characteristics was developed (in order to minimize initial mass in Earth orbit). These engine values are listed below in Table 8.

Days	80	100	150	200
Thrust, lb	28900	19380	18200	17150
Thrust, N	128550	86200	80950	76280
Specific impulse, sec	5180	4840	4800	4750
Reactor power, Mw	4520	2830	2640	2460
Initial Mass in Earth Orbit, kg	$0.94x10^6$	$0.64x10^6$	$0.38x10^6$	$0.31x10^6$
H_2 mass, kg	$0.662x10^6$	$0.428x10^6$	$0.220x10^6$	$0.164x10^6$
Engine running time, hr	72.7	65.4	35.9	27.8
Chamber temperature, °K	29760	27680	27430	27170
Chamber pressure, atm	1000	1000	1000	1000
Chamber pressure, $\frac{N}{m^2}$	$1.01x10^8$	$1.01x10^8$	$1.01x10^8$	$1.01x10^8$
Exit temperature, °K	3310	3160	3150	3135
Exit Mach No.	10.03	9.96	9.93	9.90
Exit velocity, $\frac{m}{sec}$	$5.52x10^4$	$5.15x10^4$	$5.11x10^4$	$5.06x10^4$
Exit molecular weight	2.00	2.01	2.01	2.01
Ratio of specific heats	1.255	1.263	1.263	1.263

Table 8 - Mission Characteristics for Four Mars Missions

Calculation of fission fragment formation:

"The number of fission fragments formed is calculated using the following equation:

Reactor Power (w) = fissions per sec / $3.1x10^{10}$

since the reactor powers and engine running times are known from Table 8, the number of fission fragments produced are known. It is assumed that the average molecular weight of the fission fragments is 177.5 or half the molecular weight of the U_{235} fuel. In addition, it is assumed that the number fraction of fission fragments in any unit volume is constant throughout the plume." (REF 9).

Calculation of Plume Density:

In order to calculate the number of fission fragments at any point within the plume volume, the density throughout the plume must be known. We have from REF 9 that the density far from the exit nozzle can closely be approximated by:

$$\rho = \frac{4\rho_e M_e B}{[1 + \frac{(\gamma-1)}{2M_e^2}]^{.5} \cdot \frac{2}{(\gamma+1)^{\frac{(\gamma+1)}{2(\gamma-1)}}} (\frac{r_e}{r})^2 e^{-\lambda^2(1-\cos\theta)^2}}$$

1. Where the coordinates r and θ are shown in Figure 21 below.

2. ρ is the density at any point in the plume.

3. r_e is the exit radius.

4. ρ_e is the exit density.

5. M_e is the exit Mach number.

6. γ is the ratio of specific heats.

7. B and λ are constants defined as:

$$B = \frac{(\frac{\lambda}{4\pi^{.5}})[\frac{(\gamma-1)}{(\gamma+1)}]^{.5}}{2(\gamma+1)^{\frac{1}{\gamma-1}}}$$

$$\lambda = \frac{1}{\pi^{.5}(1 - \frac{C_F}{C_{FMAX}})}$$

where C_F and C_{FMAX} are thrust coefficients.

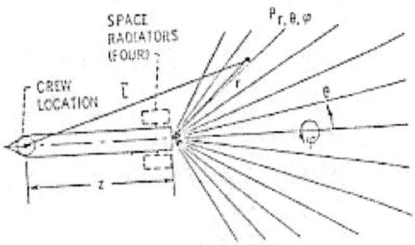

Figure 21 Relationship Between Crew Location and Fission Fragments

The radiation dose to the crew is calculated using the following from REF 9:

Radiation dose rate $\frac{rem}{hr-cm^3}$ $= 5.2x10^3 \frac{CE}{(\vec{L}\cdot\vec{L})}$

where C is the source strength at the point $P_{r,\theta,\phi}$ in the plume, in curies per unit volume, E is the photon energy in Mev, and L is the vector distance in centimeters from the point $P_{\rho,\theta,\phi}$ to the crew.

The radiation dose rate per unit volume from $P_{r,\theta,\phi}$ in rem per hour per cm^3 is:

$$\frac{5.2x10^3 F(t)}{3.7x10^{10}(\vec{L}\cdot\vec{L})} \frac{1}{2} \left(\frac{\rho A_0 P_{FF}}{m} \right)$$

Where 1 $curie = 3.7x10^{10}$ disintegrations per second.

1. A_0 is avogadros number.

2. P_{FF} is the number of fission fragments at point $P_{r,\theta,\phi}$.

3. m is the molecular weight of the nozzle exit gas.

The value of r over which the radiation density was calculated in REF 9 was expanded until a tenfold increase did not increase the radiation dose rate by more than 1%. The resultant value of r obtained was 100 km. Figure 22 below shows this relationship.

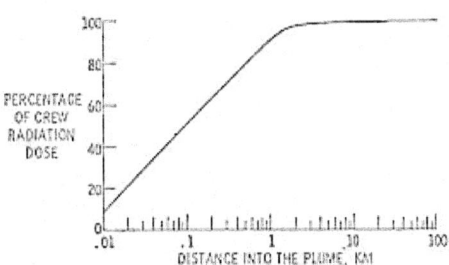

Figure 22 Crew Radiation Dose Rate vs. Integrated Distance into Plume

At a distance of 100 km essentially 100% of the gamma radiation dose to the crew is included.

Another variable of importance is the retention time of the fission fragments in the reactor core. The longer they stay in the reactor, the less of a radiation source they are to the crew. Retention times varied in the analysis from 10 sec to 10,000 sec. "The large range used for the fission fragment retention times are based on the flow system itself. The average retention time of a uranium atom is calculated to be 1000 sec, and the average retention time of a hydrogen atom is 10 sec. From this and fluid dynamic considerations, 100 sec appears to be a reasonable retention time for the fission fragments." (REF 9)

The results of the radiation dose rate for the engine associated with each mission is shown in Figure 23. One can see that the dose rate falls as retention time of the fission fragments increases from 10 sec to 10,000 sec.

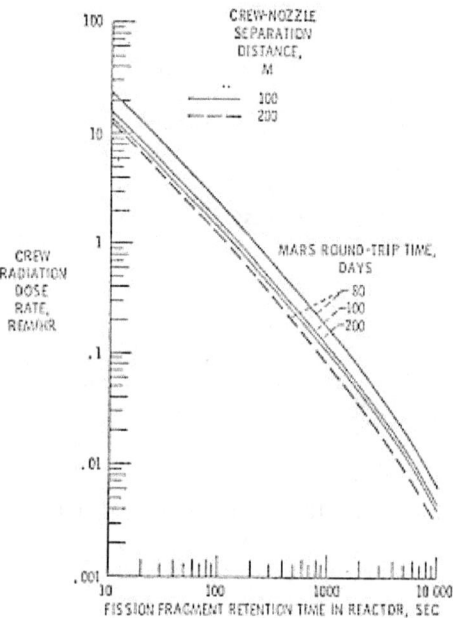

Figure 23 Fission Fragment Retention Time in Reactor vs. Crew Radiation Dose Rate.

A more important factor than dose rate is the total dose. In Figure 24, one can see the results of the total dose to the crew for the various round-trip times to Mars.

This is the unshielded case. For the 80 day round-trip time, the radiation dose is as high as 1670 rem for a fission fragment retention time of 10 sec. The radiation dose, however, drops rapidly with increasing retention time; at 10,000 sec. the radiation dose is only 0.5 rem. As the trip time increases, less energy is required and the crew dose decreases. For the 200 day round-trip, the total dose for fission fragment retention times of 10 and 10,000 sec are only 380 and 0.1 rem, respectively.

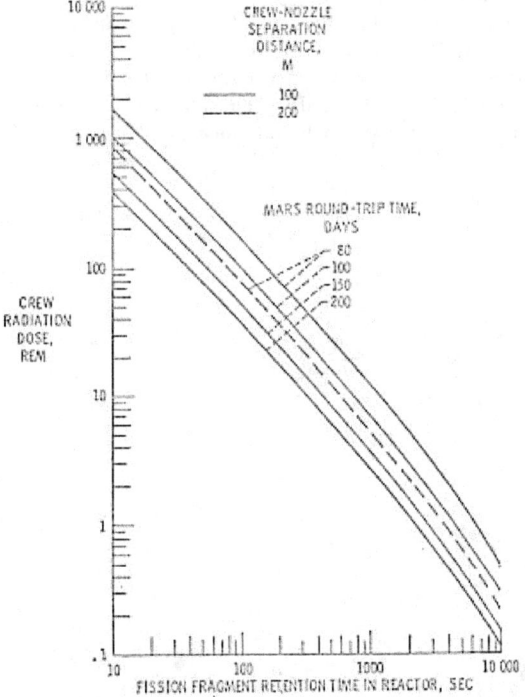

Figure 24 Fission Fragment Retention Time in Reactor vs. Total Crew Radiation Dose.

"At the most probable retention time of 100 sec, the radiation dose varies from 170 rem to 38 rem for the 80 and 200 day round-trip time, respectively. In this case the crew must be protected from the radiation dose. Five centimeters of lead shielding would reduce the radiation dose by two orders of magnitude thereby protecting the crew. The increase in vehicle weight would be insignificant. for example, a shadow shield of five centimeters thickness and four meters in diameter would add 7120 kg to the vehicle gross weight of 0.94 million kg. The equivalent attenuation of 5 cm of lead is also provided by 7.3 meters of liquid hydrogen. The amount of liquid hydrogen that is necessary for the 80 day and 200 day trips is 9500 and 2300 m^3, respectively. This is equivalent to a tank that is 10 m in diameter and 121 m long for the 80 day trip, or one that is 10 m in diameter and 29.3 m long for the 200 day trip. Therefore sufficient attenuation by this liquid hydrogen is possible during the initial portion of the trip. Also, additional attenuation is available in the form of spacecraft structure, nuclear fuel, equipment, and stores." (REF 9).

4.0   VORTEX CONTAINMENT

In order to realize the potential higher performance of the Gas-Core Nuclear Rocket it is necessary to minimize the loss rate of the gaseous nuclear fuel. One possible way of doing this is described as follows:

The principles of vortex containment for nuclear fuel in the particulate or colloidial form have been established in section 2.1. There, particulate sizes on the order of $0.5 - 5.0\mu$ were kept separate from the hydrogen propellant by means of centrifugal forces. Also, a final clean out stage was proposed in which the fuel particles in the exhaust gas could be centrifuged out from the propellant just prior to exiting the rocket via the nozzle. An equilvalent process is needed for the coaxial flow nuclear engine.

4.1   MHD APPROACH

As a small percentage of entrained nuclear fuel enters the throat of the exhaust nozzle it should be possible to separate the nuclear fuel from the hydrogen propellant by use of a MHD Driven vortex. In REF 10 a MHD-Driven rotating flow was used to simulate a gas core nuclear rocket, only in this case the idea was to have the heavy nuclear fuel gas retained near the wall of a cylindrical cavity by means of the centrifugal force produced by a rotating gas motion. The major portion of the light propellant gas would then flow axially along the centerline of the cavity, where it would be heated by radiation from the fuel region. Some of the results of this study are as follows:

> "From past studies on vortex flows, it was noted that the most effective separation of a light and heavy gas occurs in the portion of the flow that resembles solid-body rotation. Consequently, it would appear that the present scheme would perform best with a flowfield that possesses solid-body-like rotation. One possible means of producing a flow that approaches this condition is to drive the rotating gas by a Lorentz force which acts on the gas in the vicinity of the cylindrical wall of the chamber. To obtain a sharp separation between gases, the rotating flow must not only reach a high velocity but must also be free of excessive turbulence. It is believed this can best be achieved if the electrical current is spread uniformly through a volume in the shape of a cylindrical shell rather than taking the form of a single spoke."

The objective of the work in REF 10 was to produce a high speed solid-body rotation by electromagnetic means in a simulated gas core nuclear rocket chamber and explore the effectiveness of that flow in achieving a separation between heavy and light gas species.

Chamber Design:

The chamber design employed in REF 10 had the electrical current oriented in the axial direction and the magnetic field in the radial direction as shown in Figure 25 below.

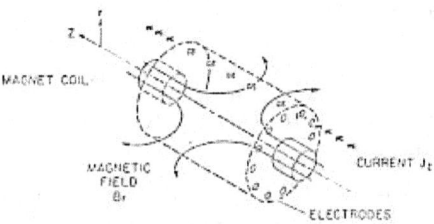

Figure 25 Current and Magnetic Field Arrangement in Experimental Apparatus

"Sixteen pairs of thoriated tungsten electrodes, each connect in series with a 2 ohm ballast resistor, were located in a circumferential pattern in the ends of the cylindrical test chamber as shown in Figure 26 below." (REF 10).

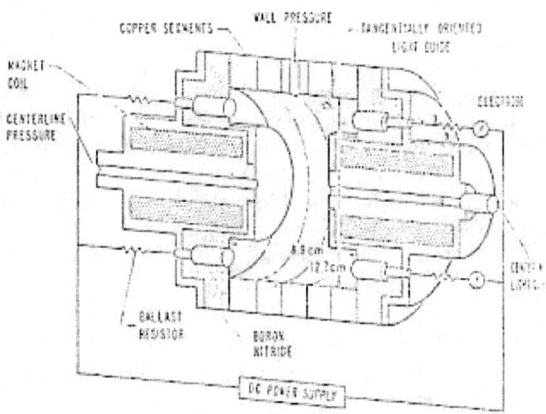

Figure 26 Chamber Schematic

"The radial magnetic field was produced by two 400-turn coils of opposing polarity aligned with the axis of the test chamber, one coil being positioned at each end of the chamber. The diameter of the coils was chosen such that the radial magnetic field increases in strength to a maximum at a radius less than that of the electrode location and then decreases towards the cylinder wall.

The cylindrical chamber encloses a volume of approximately one liter. The wall is constructed of copper rings electrically insulated from each other by boron-nitride disks. the cylinder ends are made of boron-nitride and serve as a protection for the magnet coils that protrude into the cylinder a distance of 1.6 cm as shown in Figure 26. The chamber is sealed and operates at pressures less than atmospheric. A mixture of xenon and helium is used to simulate the uranium-hydrogen mixture of a gas-core reactor. The helium is seeded with 4% hydrogen by volume for the purpose of spectroscopic diagnosis. The chamber is of a heat sink design with no cooling provided for the cylinder walls, magnet coils, or electrodes. Consequently, the tests were of short duration, typically less than 1 second." (REF 10).

Method of Operation:

"Prior to performing a test, the chamber is first evacuated and is then filled with the mixture of xenon and hydrogen-seeded helium. For most of the tests, the gas composition consisted of 56 mole percent xenon, 42.3 mole percent helium, and 1.7 mole percent hydrogen with an initial gas pressure in the chamber of either 12 or 16 torr. The operation of the device occurred in two steps. First, the discharge is initiated by drawing current from a constant-current D.C. power supply for a period of approximately 0.25 sec. At this point, while maintaining the same discharge current, the magnet is turned on to produce the (J X B) force for a duration of 0.25 to 0.35 sec. The magnet is powered by a separate D.C. supply. Although the test were of relatively short duration, it was found that a pseudo-steady-state condition was reached during both the discharge-current-only period, i.e., the J-only period, and the (J X B) period. Total discharge current varied between 400 and 2000 amps, and the radial component of the magnetic field at the maximum point between the centerline and the electrode location was in the range of 0.09-0.24 Tesla, the field strength at the electrode location being approximately 85% of the maximum. The maximum power input to the gas that was encountered in these tests was approximately 200 kw." (REF 10).

Experimental Results; Current Path:

"As mentioned earlier, the effectiveness of the Lorentz force in driving the gas and the likelihood that solid-body-like rotation will be produced are dependent on the current path. On the bases of the current measurements for individual electrodes, it was found, that when the magnetic field was applied, the current would distribute evenly among the 16 electrodes.

The path of the electrical current through the gas within the cylinder was estimated from observations of the gas luminosity. Figure 27 gives relative spectral line intensities from gas as measured along the centerline of the chamber and near the wall of the chamber." (REF 10).

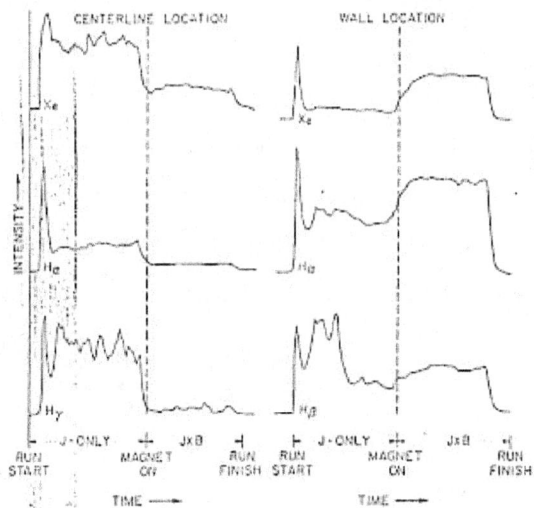

Figure 27 Spectral Line Intensities:

63

"The spectral line shown in the figure demonstrate the basic response features that are typical of most of the runs. Considering the response of the xenon line, it can be noted that the intensity is high at the center-line during the current-only portion of the run but decreases when the magnetic field is applied. Conversely, the intensity at the wall location, except for the peak at the start, is initially low but then increases as the magnetic field is applied."

Based on the results from the current measurements to individual electrodes and the behavior of the luminosity measurements, it was concluded that a reasonably even current distribution around the outer circumference of the chamber is present whenever the magnetic field is applied." (REF 10).

Gas Rotation:

One of the primary objectives of this study (REF 10) was to produce a substantial rotational motion of the gas through the use of the Lorentz force. In the experiments carried out, the rotation of the gas was evaluated by measuring the pressure at the cylinder wall and at the centerline of the chamber. The pressure distribution within the chamber is related to the rotational velocity through:

$$\rho \frac{v^2}{r} = \frac{dP}{dr}$$

where:

v = velocity

ρ = density

P = pressure

r = radial distance measured from centerline of chamber.

Thus, when rotational motion of the gas is present there will be an increase in the wall pressure and a decrease in the centerline pressure. This behavior was observed in the present tests as shown by the pressure measurements in Figure 28 below.

"The measured values of the wall-to-centerline pressure obtained from the present tests are plotted in figure 28 as a function of the product of the magnetic field and the total discharge current which is proportional to the Lorentz force. As seen from the figure, the pressure ratio seems to increase steadily with the Lorentz force, the maximum value obtained being about 2.9. As indicated in the test data shown in Table 9, the temperature near the wall is found to be higher than that at the center

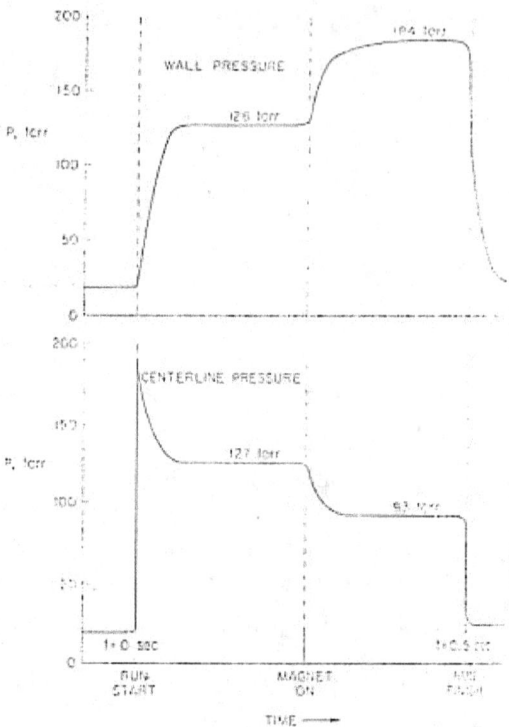

Figure 28 Pressure - Time Histories:

by approximately 20%. Under such a condition, the Mach number of the mixture corresponding to the maximum pressure ratio obtained is approximately 1.18, the Mach number for the xenon gas being about 1.35." (REF 10).

Test Conditions	
Concentration ratio in chamber before test, Xe to He	1.3:1.0
Total pressure before test	12 Torr
Total electric current	1340 Amp
Voltage gradient	$800 \frac{V}{m}$
Total arc power	134 kW
Friction loss	31 kW
Radial magnetic field strength at electrode location	0.2 Tesla
Pressure at cylinder wall during operation	133 Torr
Pressure at cylinder centerline during operation	58 Torr
State properties in centerline region	
Temperature	$8200° K$
Enthalpy	$3100 \frac{J}{gram}$
Degree of ionization of Xe	0.09
Electron density	$0.5 \times 10^{15} cm^{-3}$
State properties in arc region	
Temperature	$10,200° K$
Enthalpy	$4800 \frac{J}{gram}$
Degree of ionization of Xe	0.44
Electron density	$2.8 \times 10^{16} cm^{-3}$
Frozen sound speed for mixture	$1.5 \frac{km}{sec}$
(Frozen) Mach number for mixture	0.86
(Frozen) Mach number for Xe	1.05
Tangential velocity	$1.29 \frac{km}{sec}$
Electrical conductivity	$3000 \frac{mho}{m}$
Hall parameter	0.104
Reynolds number based on diameter	1.1×10^{5}

Table 9 - MHD Test Data

Prototype rocket:

Appling these results to our requirement for a MHD driven vortex in the exit nozzle of the coaxial flow engine; we have from standard nozzle design that the Mach number in the throat will be unity, therefore just before the throat of the nozzle the Mach number will be in the range of 1.1 to 1.4 for the fuel propellant mixture. These values correspond with the values of Mach number obtained from REF 10. However, it was stated in REF 10 that the results from there tests could not be compared directly with the conditions present in the prototype nuclear rocket since a mixture heavily enriched with the heavy gas was used in their tests instead of a light-gas-enriched mixture that is more suited to the prototype. Such a deviation from the exact simulation was made in REF 10 in the interest of maintaining a relatively high electrical conductivity (which would result from the fissioning process in an actual gaseous core reactor).

Also, in a prototype, the fuel gas will be multiply ionized and thus will behave as a uniform conducting medium. Because of this difference, it was stated in REF 10 that the problems associated with the current discharge will probably be less in an actual gas-core rocket. Thus, the possible adverse effect of increased pressure on the performance would diminish in a rocket.

5.0   NUCLEAR CRITICALITY

While the gas-core reactor provides improved performance over the solid core reactor, it suffers from large size and weight requirements. One method of reducing these requirements is to combine the high performance (specific impulse) of the gas core with the compactness and low weight of a NERVA type solid core reactor.

"This concept has a central cavity region surrounded by a moderator region which thermalizes neutrons from a driver region located outside of the moderator region but inside of a pressure vessel. The driver region can use fuel elements developed in the NERVA program and can be operated with an enert gas such as argon in place of hydrogen for cooling." (REF 11).

In REF 11 a cavity diameter of 0.61 m was selected. The reactor regions for the reflector and driver(fuel) elements were allowed to vary, but the overall diameter of 1.22 m as shown in Figure 29 was maintained.

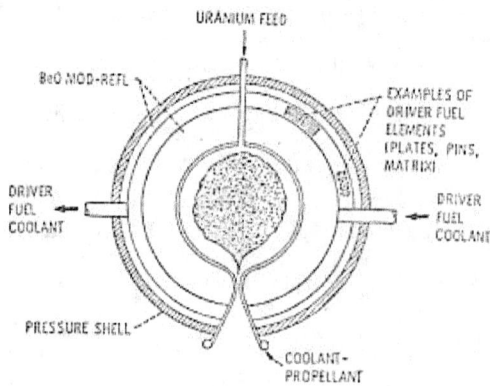

Figure 29 Driver-Gas-Core Reactor Concept

"The uranium fuel region in the cavity was held to a diameter of 0.42 m ($\frac{D_F}{D_C}$). The outermost region of the Driver-Gas-Core reactor (also known as the Mini-Cavity reactor) is the pressure vessel. Because the driver region is located near the outer periphery of the reactor there is considerable fast neutron leakage. By adding a thick fast reflector such as the pressure vessel, the reactivity is enhanced by the reflection of the fast neutrons back into the reactor, thus allowing a reduction of fuel density in the driver region". (REF 11.)

The important limitation is that the uranium in the solid fuel driver region does not become hot enough to vaporize. "The pressure shell material was an annealed titanium alloy ($T_i, 6AL, 4U$) because of it's strength to weight ratio, high strength at high pressures, and established fabricability. The radiator used consisted of beryllium fins and T2M(molybdenum alloy) tubes. The temperature for the radiator was taken as 1000°C." (REF 11).

Reactor Power splits between Driver and Cavity Regions:

The power splits (cavity and driver) are presented in Table 10 for a range of thrust and pressure levels. also presented in the table is the mass required in the driver region for criticality.

Thrust N (lbf)	Pres. atm	Propellant flow rate kg/sec	Fuel (cavity) mass, (kg)	Cavity power MW	Driver power MW	Fuel(driver) mass kg	Fuel(cavity) edge temp °C	Fuel vap temp °C cham
222 (50)	200	0.020	0.97	1.67	9.0	23	5000	6690
	500	0.016	1.79	1.97	7.2	18	6400	7525
	1000	0.014	2.80	2.42	6.0	13	8050	8325
445 (100)	200	0.035	0.76	3.95	24.2	24	6100	6690
	500	0.028	1.42	4.55	19.7	20	7800	7525
	1000	0.025	2.24	5.46	16.7	15	9150	8325
668 (150)	200	0.047	0.66	6.64	43.4	25	6400	6690
	500	0.039	1.23	7.64	36.0	21	8300	7525
	1000	0.035	1.96	8.91	30.5	17	10500	8325
890 (200)	200	0.060	0.60	9.28	62.1	26	6700	6690
	500	0.049	1.11	10.90	54.7	22	9100	7525
	1000	0.044	1.79	12.50	45.9	18	11675	8325

Table 10 Reactor Power Splits and Fuel Loading

The powerplant weight consists of weights for the reactor, the pressure shell, the pumps, and the radiator. The reactor weights approximately 2180 kg. The weight of the pumps is estimated to be 2180 kg. The weights of the pressure shell and radiator were allowed to vary with reactor pressure and power. Powerplant characteristics and weight are shown in Table 11 below.

Thrust N (lbf)	Pressure (atm)	Specific impulse sec	Propellant flow rate, kg/sec	Driver power, MW	Reactor wght. lb	Pressure shell wght. lb	Radiator wght., lb	Total Power-plant wght. lb
222	200	1100	0.020	9.0	4800	800	4350	10150
(50)	500	1375	0.016	7.2	4800	1860	5780	12640
	1000	1600	0.014	6.0	4800	4000	7880	16880
445	200	1300	0.035	24.2	4800	800	11700	17500
(100)	500	1600	0.028	19.7	4800	1860	15820	22680
	1000	1800	0.125	16.7	4800	4000	22200	31200
668	200	1460	0.047	43.4	4800	800	21000	26800
(150)	500	1770	0.039	36.0	4800	1860	28900	35760
	1000	1960	0.035	30.5	4800	4000	40590	49590
890	200	1530	0.060	62.1	4800	800	30060	35860
(200)	500	1850	0.049	54.7	4800	1860	43900	50760
	1000	2060	0.044	45.9	4800	4000	61100	70100

Table 11 Powerplant Characteristics and Weight

6.0   PROPOSED GAS-CORE DESIGN

The goal of the present Gas-Core Nuclear Rocket design is to provide a means of interplanetary travel which does not require inordinate amounts of travel time, hundreds of billions of dollars of expense, or massive energy expenditures required in lifting large amounts of mass into initial low earth orbit. With these considerations in mind the following design criterion were adopted.

1. Of the (3) Gas-Core designs reviewed (Colloid-Core, Nuclear Light Bulb, and Coaxial Flow) the Coaxial Flow Gas-Core Nuclear Rocket Engine was chosen as the base design to build upon. This was because the Coaxial Flow has the highest performance potential of the (3), with specific impulse up to 5000 seconds.

2. In order to reduce the nuclear fuel losses from the gas core a MHD choke (stage) should be added just prior to the nozzle throat, where the Mach numbers are in the range of 1.5 to 1.0. There, nuclear fuel entrained with the propellant will be centrifugally separated and returned to the central fuel region of the cavity.

3. In order to reduce weight and high pressure requirements, driven by criticality needs, a Nuclear Driver Region should be added which can supply fast neutrons to the fuel cavity region.

4. Since reduced trip times are a goal (80 Days to Mars) radiation levels will pose a danger to the crew, so a radiation shield may be required.

5. Hydrogen is obviously the best propellant to use, and maintaining it in the Slush Hydrogen State will provide significant savings in tankage volume and weight.

6. Since extremely high temperatures will be reached in the cavity and nozzle areas, transpiration cooling will be required in the nozzle and space radiators will be required for the cavity cooling system.

6.1   GAS-CORE DESIGN DETAILS

The final design chosen is based in part (size and weight of gas core chamber) on the design of Table 6, page 38. Engine thrust is 250,000 lbs. and structural weight is about 100,000 lbs (45,360 kg). However, for this design the Specific Impulse required will be 5000 seconds. From Table 8, page 48 for a 5000 second I_{sp} Coaxial-Flow Nuclear Rocket, internal chamber pressures are approximately 1000 atmospheres and propellant temperatures are about $27,000°K$.

Specific Impulse can be converted to exhaust velocity by multiplying by g_e thus:

$$U_e = (5000secs)x(9.8\frac{m}{s^2}) = 49,000(\frac{m}{s})$$

or:

$$U_e = 49(\frac{km}{sec})$$

now:

$$(1)lbf = (.4536(kg))x(9.8(\frac{m}{s^2})) = 4.445(\frac{kg-m}{s^2})$$

or: $4.445N$

so:

$$Thrust = (250,000lbf)x(4.445\frac{N}{lbf}) = 1.11x10^6(N)$$

now the Mass Flow is in (lbm/sec), and I_{sp} is in: $I_{sp} = \frac{lbf}{(\frac{lbm-lbs}{sec})}$

or:
$$\frac{lbm}{sec} = \frac{lbf}{g_c - I_{sp}} = \frac{lbf}{U_e}$$

so:
$$\text{Mass Flow} = \frac{1.11 \times 10^6 \left(\frac{kg - m}{s^2} \right)}{49,000 \left(\frac{m}{s} \right)}$$

or:
$$\text{Mass Flow} = 22.65 \left(\frac{kg}{s} \right) \text{ Hydrogen Propellant}$$

Now we have:

1. $I_{sp} = 5000$ secs.

2. $Thrust = 1.11 \times 10^6$ Newtons.

3. $U_e = 49 \left(\frac{km}{s} \right)$

4. $\dot{m} = 22.65 \left(\frac{kg}{s} \right)$

   and from REF 6:

5. Cavity Pressure = 1000 atm.

6. Maximum Fuel Temperature = $120,000^\circ R$, or $66,000^\circ K$

7. Average Fuel Temperature = $90,000^\circ R$, or $50,000^\circ K$

8. Average Propellant Temperature = $10,000^\circ R$ to $20,000^\circ R$ or, $5,500^\circ K$ to $11,000^\circ K$.

9. Amount of Seed Material < 1% Hydrogen Mass Flow (negligible effect on Specific Impulse).

6.2   GAS CORE DESIGN DETAILS-MISSION ANALYSIS

In order to further refine our design the intended mission must be outlined. To this
end I will calculate the requirements for interplanetary missions to Mars & Saturn.
These two goals were chosen in the hopes of developing a vehicle which could perform
missions to both the inner and outer solar system.

Starting with the Vis-Viva equation (REF 12):

$$\frac{ds}{dt}^2 = k^2(m_1 + m_2)(\frac{2}{r} - \frac{1}{a})$$

where:

1. $\frac{ds}{dt}$ is the speed of the orbiting body.

2. m_1 is the mass of the major body.

3. m_2 is the mass of the orbiting body.

4. k is the gravitational constant.

5. a is the semi-major axis of the orbit

6. r is the distance of the orbiting object from the force center.

From this we can derive the earth's orbital speed(with $a = 1$):

$$\frac{ds}{dt}^2 = k_s^2(m_1 + m_2)(\frac{2}{r})$$

Now m_1 = the mass of the Sun and m_2 = the mass of the earth, so $m_1 \gg m_2$,
therefore:

$$\left(\frac{ds}{dt}^2\right) = 2.959122x10^{-4}(1+0)\frac{2}{2}$$

$$\frac{ds}{dt} = (0.017202)\left(\frac{AU}{day}\right)$$

$$\frac{ds}{dt} = \frac{(0.017202)(149.6x10^9)}{(24)(60)(60)} = 29,785\left(\frac{m}{s}\right)$$

For Mars $a = 1.52$ and a similar calculation yields:

$$\frac{ds}{dt} = 24,157\left(\frac{m}{s}\right)$$

And for Saturn:

$$\frac{ds}{dt_{Saturn}} = 9,600\left(\frac{m}{s}\right)$$

Let $\mu = m_1 + m_2$ where m_1 is the mass of the Sun in solar masses and m_2 is the mass of the space vehicle in the same units. Clearly $m_1 \gg m_2$ and so $\mu = 1$.

And let $\tau = k_s(t - t_0)$ then the Vis-Viva equation reduces to:

$$\dot{S}^2 \stackrel{def}{=} \left(\frac{ds}{dt}\right)^2 = \left(\frac{ds}{k_s dt}\right)^2 = \mu\left(\frac{2}{r} - \frac{1}{a}\right)$$

At the assumed Earth circular orbit $r = 1$ AU and our equation reduces to:

$$\dot{S}^2 = (1)\left(\frac{2}{a} - \frac{1}{a}\right) = (1)\frac{1}{a} = 1$$

hence our speed from the above equation is consistent with the orbital speed at unit distance, ie.

Distance= 1 AU

Earth Mean Orbital Speed (E.M.O.S.)= 1 (In terms of EMOS $V = 29.766(\frac{km}{sec})$)

Let us calculate the Delta S required to transfer to Mars.

$$A_{Transfer} = (\frac{a_{earth} + a_{mars}}{2}) = \frac{(1 + 1.52)}{2} = 1.26(AU)$$

At injection to the transfer orbit:

$r = 1$ and so:

$$\dot{S}_1^2 = \mu(\frac{2}{r} - \frac{1}{a}) = (1)(\frac{2}{1} - \frac{1}{1.26})$$

$$\dot{S}_1 = 1.0983$$

Of course the Earth is traveling around the Sun at unit speed, so that the actual speed increment that needs to be added to the space vehicle is $1.0983 - 1.00 = .0983$ or in terms of $\frac{km}{sec}$:

$$(.0983)(29.766) = 2.926(\frac{km}{sec})$$

At aphelion on the transfer ellipse (that is at the point of injection of the payload onto the destination orbit),

80

$$r = 1.52$$

so:

$$\dot{S}_2^2 = \mu(\frac{2}{r} - \frac{1}{a}) = (1)(\frac{2}{1.52} - \frac{1}{1.26})$$

$$\dot{S}_2 = .72259$$

The speed required to establish the payload on the orbit of Mars at 1.52 AU is the difference between $\dot{S}_2$ and the circular orbit speed at 1.52 AU, which is:.

$$\dot{S}_3^2 = (1)(\frac{2}{r} - \frac{1}{a}) = (1)(\frac{2}{1.52} - \frac{1}{1.52})$$

$$\dot{S}_3^2 = \frac{1}{1.52}$$

or:

$$\dot{S}_3 = \frac{1}{\sqrt{1.52}}$$

Thus, the final injection speed required to change from the transfer orbit to the Mars orbit is:

$$\dot{S}_3 - \dot{S}_2 = \frac{1}{\sqrt{1.52}} - .72259 = .0885$$

Or in terms of $\frac{km}{sec}$:

$$(.0885)(29.766) = 2.634(\frac{km}{sec})$$

The Delta V or total speed change that must be accomplished by the rocket is:

$$\dot{S}_1 - (\dot{S}_3 - \dot{S}_2) = .0983 + .0885 = .1868$$

Presumably our travelers would eventually like to come home so our total requirement is:

$$\Delta S = 2(.1868) = .3736$$

or:

$$(.3736)(29.766) = 11.120(\frac{km}{sec})$$

For a mission to Saturn the velocity requirements are:

$$\dot{S}_1^2 = (1)(\frac{2}{1} - \frac{1}{5.25})$$

$$\dot{S}_1 = 1.34518$$

and taking into account the Earth's velocity of 1.00

$$\dot{S}_1 = 1.34518 - 1.00 = .34518$$

at $r = 9.5$ AU

$$\dot{S}_2^2 = (1)(\frac{2}{9.5} - \frac{1}{5.25})$$

$$\dot{S}_2 = .14159$$

$$\dot{S}_3^2 = (1)(\frac{2}{9.5} - \frac{1}{9.5}) = \frac{1}{9.5} = .10526$$

$$\dot{S}_3 = .32444$$

$$\dot{S}_3 - \dot{S}_2 = .32444 - .14159 = .18285$$

$$\Delta V = \dot{S}_1 + (\dot{S}_3 - \dot{S}_2) = .34518 + .18285$$

$$\Delta V = .52803$$

Considering the return trip:

$$\Delta V = (.52803)x2$$

$$\Delta V = 1.05606$$

$$\Delta V = (1.05606)x(29.766) = 31.434(\frac{km}{sec})$$

To compute the mass of slush hydrogen propellant required to accomplish this trip we make use of the famous rocket equation for a single stage, in the absence of a gravity field (ie. effects of Earth & Mars gravity not taken into account). (REF 13).

$$\Delta u = u_e \ln(\frac{M_i}{M_b})$$

Where:

M_i =Initial Mass of Rocket (includes propellant mass, structure mass, and payload mass).

M_b =Burnout mass (in this case payload mass and structure mass).

u_e = Propellant Exhaust Velocity

We also have:

$$M_b = M_p + M_s$$

Where:

M_p = Payload Mass

M_s = Structure Mass

and:

$$M_s(engine) = 45,360 \text{ kg}$$

Lets assume that $M_p + M_s(tankage) = 45,360$ kg

Then:

$$M_i = M_b e^{\left(\frac{\Delta V}{u_e}\right)}$$

Since or Saturn mission requires:

$\Delta V = 31.434\left(\frac{km}{sec}\right)$ minimum, lets set:

$$\Delta V \stackrel{\text{def}}{=} 35\left(\frac{km}{s}\right)$$

Then:

$$M_i = (45,360x2)e^{\left(\frac{35}{49}\right)} = 185,316(kg)$$

For a mass ratio of:

$$\frac{M_i}{M_b} = \frac{185,316}{(45,360x2)} = 2.04$$

The mass of propellant required is: $185,316 - (45,360x2)$

$$M_{Propellant} = 94,596(kg)$$

The total burning time is:

Propellant Mass / Propellant Mass Flow Rate $= \frac{94,596(kg)}{22.65(\frac{kg}{sec})}$

$T_{burning} = 4176$ secs or 69.6 minutes

Now lets look at our travel times:

from Keplers Law (REF 12): $P = P_0 a^{\frac{3}{2}}$

For our transfer orbit to mars:

$$a_t = \frac{1. + 1.52}{2} = 1.26\,AU$$

$$P = (1yr)(1.26)^{\frac{3}{2}} = 1.414yrs$$

and the time to complete a half orbit is: $(\frac{1.414}{2}) = .707yrs$

and for Saturn $a_t = 5.25$

$$P = (1yr)(5.25)^{\frac{3}{2}} = 12.029yrs$$

and the time required to complete a half orbit is about 6 years.

However, these calculations are based on:

1. A Mass Ratio of 2.04 to 1.

2. ΔV of 35 km/sec.

If we increase the mass ratio (i.e. the amount of propellant in relation to the structural and payload mass) to a resonable amount, say 5, then:

$5 = \frac{M_t}{M_b} = e^{\frac{\Delta V}{u_e}}$ and $\ln(5) = \frac{\Delta V}{u_e}$

$$\Delta V = u_e \ln(5) = 78.86(\frac{km}{s})$$

The new mass ratio is:

$$\frac{M_P}{(45,360 x 2)} = 5$$

and the mass of the propellant is:

$$M_p = 453,600(kg)$$

and our burning time becomes:

$$\frac{M_p}{\dot{m}_p} = \frac{453,600(kg)}{22.65(\frac{kg}{sec})} = 20,026(secs)$$

$= 5.56$ hrs. total burning time.

This time will be mainly consumed during the four (4) burn periods; injection upon the transfer orbit, injection upon the destination orbit, injection upon return transfer orbit, and injection upon circular orbit at earth's distance from the Sun.

The previous travel times of .707 and 6 years for trips to Mars and Saturn respectively, were based on minimum energy transfer orbits. Our 5000 sec I_{sp} nuclear rocket can do much better. Lets set our $\dot{S}_1$ at 48 km/s; that is 30 km/s due to the Earth's orbital speed and 18 km/s added by our rocket. Then; we insert this value into the Vis-Viva equation:

$$\dot{S}_1^2 = \mu(\frac{2}{r} - \frac{1}{a})$$

with:

$$\mu = 1$$

and

$$r = 1$$

$$-\frac{1}{a} = \dot{S}_1^2 - 2 = (\frac{48}{30})^2 - 2$$

$$-\frac{1}{a} = (1.6)^2 - 2 = .56$$

and $a = -1.785$ since 'a' is negative, we will be following a hyperbolic course instead of an ellipse.

At Mars:

90

$$\dot{S}_2^2 = (1)(\frac{2}{1.52} - \frac{1}{-1.785}) = 1.876$$

and:

$$\dot{S}_2 = 1.3696$$

The orbital speed of Mars is:

$$\dot{S}_3 = \frac{1}{\sqrt{1.52}} = .811$$

Since this is less than our spacecraft speed we must slow down, and the required speed change is:

$$\dot{S}_2 - \dot{S}_3 = 1.3696 - .811 = .5586$$

and:

$$\dot{S}_1 + (\dot{S}_2 - \dot{S}_3) = .6 + .5586 = 1.1586$$

and, including the return trip:

$$(2x1.1586) = 2.3172$$

or:

$$\Delta V = (2.3172)x30(\tfrac{km}{s}) = 69.5(\tfrac{km}{s})$$

Well within our performance potential of:

91

$$\Delta V = 78.86 \left(\frac{km}{s} \right)$$

The transit time is calculated using Kepler's Equation formulated for the Hyperbola (REF 12), namely:

$$\frac{k_s \sqrt{\mu}}{-a^{\frac{3}{2}}} (t - T) = e \sinh(F) - F$$

At injection time t_1 (at Earth)

$$r = a(1 - e \cosh(F_1))$$

or:

$$1 = -1.785(1 - 1 \cosh(F_1))$$

Where: $e = 1$ because this is a rectilinear hyperbola.

Solving for F_1:

$$-.560 = 1 - \cosh(F_1)$$

$$F_1 = \cosh^{-1}(1.56) = 1.014$$

At Mars:

$$1.52 = -1.785(1 - (1)\cosh(F_2))$$

$$-.8515 = 1 - \cosh(F_2)$$

$$F_2 = \cosh^{-1}(1.8515) = 1.2266$$

and:

$$(t_2 - t_1) = ((1)\sinh(F_2) - (1)\sinh(F_1) + F_2 - F_1)\frac{-a^{\frac{3}{2}}}{k_s\sqrt{\mu}}$$

$$(t_2 - t_1) = (1.558 - 1.1969 + 1.2266 - 1.014)\frac{-a^{\frac{3}{2}}}{k_s\sqrt{\mu}}$$

$$(t_2 - t_1) = (.5737)\frac{1.785^{\frac{3}{2}}}{.017,202,098}$$

$(t_2 - t_1) = 79.53$ Days or about 80 Days to Mars.

And for Saturn:

$$\dot{S}_1 = .6$$

$$\dot{S}_3 = .32444$$

$\dot{S}_2^{\,2} = (1)(\frac{2}{9.5} - \frac{1}{-1.785})$ since 'a' is the same.

$$\dot{S}_2 = .8779$$

$$\dot{S}_2 - \dot{S}_3 = (.8779 - .32444) = .55346$$

$$\dot{S}_1 - (\dot{S}_2 - \dot{S}_3) = .6 + .55346 = 1.15346$$

And including the round trip the total ΔV is:

$$\Delta V = (2 x 1.15346) = 2.3069$$

or:

$$(2.3069)(30\frac{km}{sec}) = 69.2(\frac{km}{sec})$$

The travel time to Saturn is computed as follows:

Since 'a' is the same:

$$F_1 = 1.014$$

94

At Saturn:

$$9.5 = -1.785(1 - (1)\cosh(F_2))$$

$$-5.322 = 1 - \cosh(F_2)$$

$$F_2 = \cosh^{-1}(6.322) = 2.530$$

and:

$$(t_2 - t_1) = (\sinh(2.53) - \sinh(1.014) + 2.53 - 1.014)\frac{1.785^{\frac{3}{2}}}{k_s}$$

$$= (6.2369 - 1.1969 + 2.53 - 1.014)\frac{2.3848}{k_s}$$

$$= 6.556(\frac{2.3848}{.017,202,098})$$

$(t_2 - t_1) = 908.8$ Days or 2.488 years.

To summarize we have:

1. $I_{sp} = 5000$ Sec.

2. Thrust$= 250,000$ Lbs.

3. $\dot{m} = 22.65(\frac{kg}{s})$

4. Mass Propellant $= 453,600$ kg

5. Mass Payload $= 45,360$ kg

6. Mass Structure $= 45,360$ kg

7. Mass Ratio $= 5$

8. $U_e = 49(\frac{km}{sec})$

9. $\Delta V = 78.86(\frac{km}{s})$

10. Total Burning Time $= 5.56$ Hours

6.3   GAS CORE DESIGN DETAILS - MHD CHOKE

REF (10) had solid-body-like rotation of a shell(cylindrical) near the cavity wall. What is needed here is solid-body-like rotation in the nozzle area. But not a shell, since in this case the entrained fuel is not located in the perifery, but in the center and throughout the nozzle area. However, the area in question is much smaller than the cavity area so that it should be possible to establish the required solid-body-like rotation. Note that the fuel which is about to be expelled out the nozzle must be moved all the way from the center of the nozzle area to the nozzle wall location.

If we use the value of 2.9 obtained from Figure 28 and the 1% loss rate for the uranium fuel from Table 6 for the $1.11x10^6$ N thrust Coaxial Flow engine, the new loss rate becomes:

$$\left(\frac{.01}{2.9}\right) = .0034 \text{ or } .34\%$$

Now if we take our $M_p = 453,600$ kg. the required mass of fuel becomes:

$$M_f = (453,600x.0034) = 1542.24(\text{kg})$$

6.4   GAS CORE DESIGN DETAILS - NUCLEAR CRITICALITY

Here the idea is that; "the fissioning plasma does not necessarily have to be critical by itself. If a sufficiently large external neutron flux is available, and if a fissionable gaseous fuel can be made dense enough, such that the fission fragment stopping distance becomes comparable or smaller than the dimension of the fuel volume, then the volume of the central fuel region of the cavity and the amount of nuclear fuel can be reduced." (REF 12). For example, the Los Alamos Nuclear Furnace has a test section which can provide a neutron flux density of $10^{15}(\frac{Neutrons}{cm^2-sec})$.

"A research design proposed by Rom and Ragsdale is similar to the mini-cavity design, however, with a differently shaped solid fuel driver region. The solid fuel driver section is cylindrical for easy insertion and removal of fuel elements. The reactor is heavy water and beryllium moderated, and housed in a pressure shell. Provisions are made for through-flow of propellants and the feeding of nuclear fuel into the test section. In the test section a neutron flux density of several times $10^{14}(\frac{Neutrons}{cm^2-sec})$ up to $10^{15}(\frac{Neutrons}{cm^2-sec})$ is achievable, sufficiently high to test coaxial flow configurations". (REF 12).

6.5   GAS CORE DESIGN DETAILS - RADIATION LEVELS

Using the data from Figure 30 below and scaling up the reactor power level for the 80 day Mars trip to 38,793 MW (corresponding to the increased thrust requirement of 250,000 Lbs). Also, assuming 5.56 hours total running time, which yields a total mission energy requirement of 215×10^3 Mw-h. We see that for a fission fragment retention time of 100 seconds in the cavity, the crew radiation dose is approximately 100 REM.

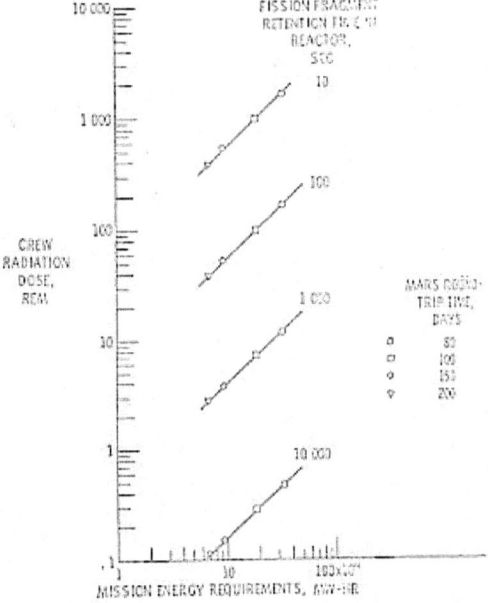

Figure 30 Mission Energy Requirements vs. Crew Radiation Dose

However, assuming a more probable fission fragment retention time of 1000 sec (with the MHD-Driven Nozzle Choke) we get a total crew radiation dose of 8 REM. As a result of this, a smaller radiation shield than the one proposed at the end of section 3 may be required.

6.6   GAS CORE DESIGN DETAILS - PROPELLANT

The use of 50% Slush Hydrogen SH_2 minimizes propellant storage volume requirements. While this requires increased launch (in orbit) and ground support, it should impose no operational difficulties.

One design for an expendable nuclear injection stage (to be launched from the Space Shuttle) is shown in Figure 31 (next page).

"The module tanking was fabricated from 2014-T651 aluminum alloy. The membrane thickness was determined based on a differential pressure of 22 PSIA to which a factor of safety of 1.4 was applied to arrive at the ultimate design pressure of 30.8 PSIA. The worst design condition is at the end of engine operation when the tank is full of hot ullage gas at a temperature of 250° R. At this temperature the ultimate tension allowable stress of the selected material was 72,000 PSI." (REF 14).

This particular tank has a capacity of 29,350 Lbs of SH_2 with a 3% ullage volume at loading.

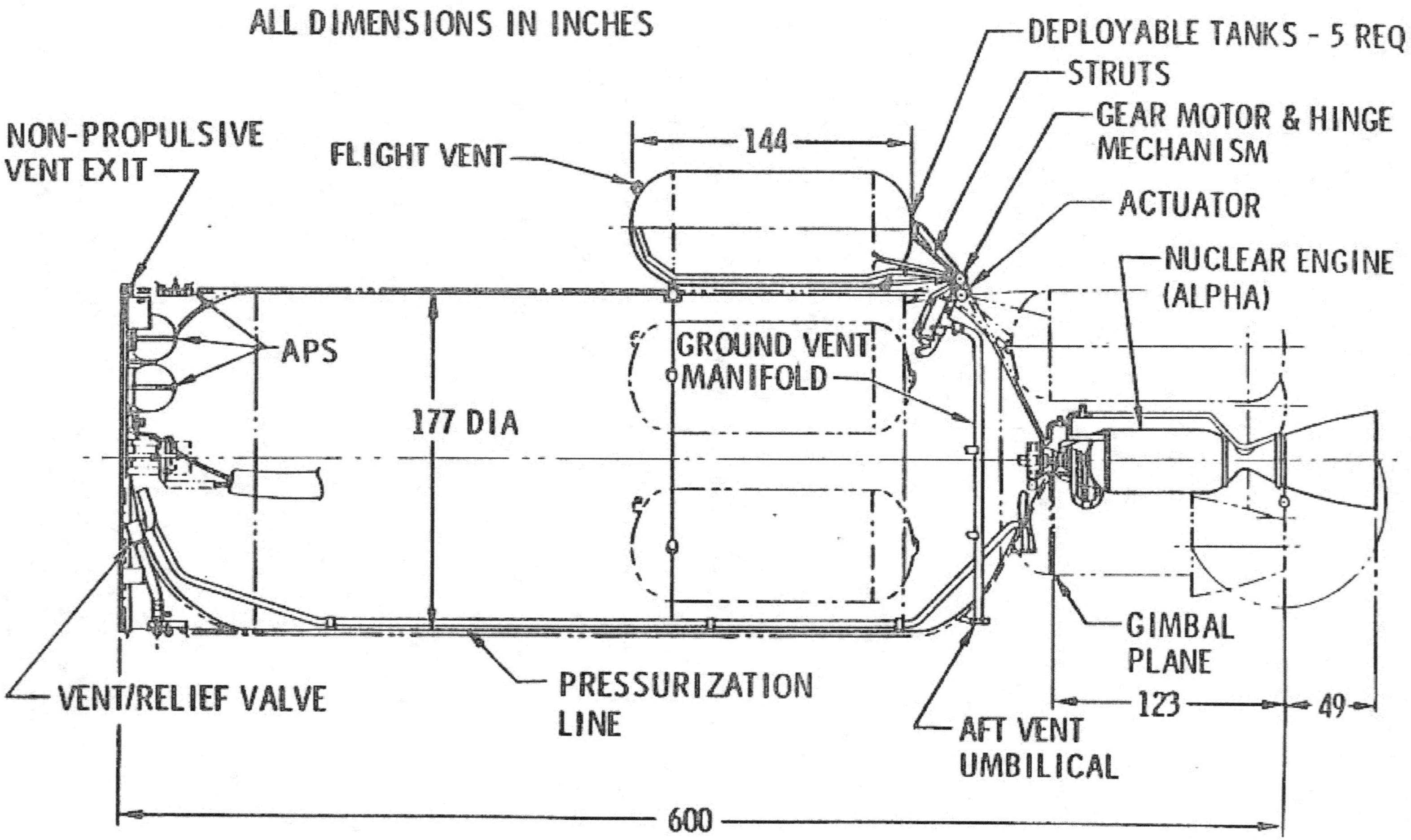

ALL DIMENSIONS IN INCHES

Figure 31 Example Slush Hydrogen Propellant Tank

102

7.0   SUMMARY AND RECOMMENDATIONS

Further research and experimentation leading to the development of the Coaxial-Flow Gas Core Nuclear Rocket should proceed along the following lines. (REF 15).

1. Demonstration of criticality of gaseous Uranium fuel in a cold static configuration.

2. Investigations of the stability of fissioning plasmas, and of the spectral distribution of radiated power.

3. The confinement of a gaseous fuel by fluid mechanic means in Cold-Flow tests.

4. Demonstration of Criticality of gaseous uranium confined in a Cold-Flow configuration.

5. Propellant heating by radiation from an electrically or fission heated gas.

6. The confinement of a hot gaseous fuel by fluid mechanic means.

7. A prototype test configuration which incorporates the above mentioned items.

NASA plasma core research as of Sept. 1972 is outlined in Table 12 below. The 'A' stands for analytical work. The 'E' indicates experimental investigations.

Criticality of Gaseous Fuel	Radiative heat Transfer	Nuclear Fuel Confinement	Systems Study
NASA Lewis (A)	NASA Lewis (A/E)	NASA Lewis (A)	NASA Lewis (A)
Aerojet Nuclear (E)	TAFA Division Humphreys Corp. (E)	Ill. Inst. of Technology (A/E)	United Aircraft Research Labs. (A)
United Aircraft Research Labs. (A)	Ga. Inst. of Technology (E)	Cornell Univ. (A/E)	Ga. Inst. of Technology (A)
Versar Inc. (A)	AEDC (E)	United Aircraft Research Labs. (A/E)	Univ. of Florida (A)
	United Aircraft Research Labs. (A/E)	Aerojet Nuclear (E)	Computer and Applied Sciences (A)
	Univ. of Florida (E)	Univ. of Arizona (E)	
	Univ. of Maryland (E)	NASA Ames (E)	
	NASA Langley (E)		

Table 12 NASA Uranium Plasma Research

Before any of these activities can continue from their suspended beginnings in 1973, manned interplanetary missions must be designated as a national goal. Judging from the decisions of our political leaders of the past decade and a half, this is not likely to occur in the near future.

One is reminded that the use of nuclear energy for rocket propulsion was first proposed by Leo Szilard in 1932. He said then that only through the liberation of atomic energy could we obtain the means which would enable man not only to leave the earth, but to leave the solar system. This is as true today as it was then.

APPENDIX A

LIST OF REFERENCES

1. An Engineering Study of the Colloid-Fueled Reactor Concept, Y.S. Tang, J.S. Stefanko, P.W. Dickson, and D.W. Drawbaugh. Journal of Spacecraft and Rockets, Vol. 8, No. 2, February 1971.

2. Two-Component Vortex Flow Studies of the Colloid Core Nuclear Rocket, Loren A. Andersen, Siegfried H. Hasinger, and B.N. Turman. Journal of Spacecraft and Rockets, Vol. 9, No. 5, May 1972.

3. Experimental Flow Studies of the Colloid Core Reactor Concept, B.N. Turman and Siegfried H. Hasinger, Journal of Spacecraft and Rockets, Vol. 9, No. 10.

4. Performance Potential of the Colloid Core Reactor Concept in Near-Earth Applications, CAPT. Thomas C. Meier (USAF). Journal of Spacecraft and Rockets, Vol. 10, No. 9, September 1973.

5. Gas-Core Nuclear Rocket Engine Technology Status, George H. McLafferty, Journal of Spacecraft and Rockets, Vol. 7, No. 12, December 1970.

6. Nuclear Rocket Propulsion, Frank E. Rom, NASA TMX-1685, November 1968

7. Buoyancy Effect on Fuel Containment in a Gas-Core Nuclear Rocket, Henry A. Putre, Journal of Spacecraft and Rockets, Vol. 9., No. 12.

8. Acoustic Instabilities in a Constant Flux Gas Core Nuclear Rocket, Harry McNeil and Martin Becker, AIAA Journal, Vol. 8, No. 2.

9. Crew Radiation Dose from a High-Impulse Gas-Core Nuclear Rocket Plume, Charles C. Masser, Journal of Spacecraft and Rockets, Vol. 9, No. 8, August 1972.

10. An Experiment on the MHD-Driven Rotating Flow for a Gas Core Nuclear Rocket, Wendell L. Love and Chul Park, AIAA Journal, Vol. 8, No. 8, August 1970.

11. A Mini-Cavity Reactor for Low-Thrust, High-Specific Impulse Propulsion, Robert E. Hyland, Journal of Spacecraft and Rockets, Vol. 9, No. 8, August 1972.

12. Astrodynamics - Applications and Advanced Topics, Robert M.L. Baker, Jr., West Coast University. 1986.

13. Mechanics and Thermodynamics of Propulsion, Hill-Peterson. Addison-Wesley Publishing, November 1970

14. Nuclear Propulsion Short Course System Design and Integration, V.E.Haloulakos, McDonnell Douglas Astronautics Co.

15. Review of Fission Engine Concepts, Karlheinz Thom, Journal of Spacecraft and Rockets, Vol. 9, No. 9, 1972.